Functional Skills Maths

In Context

ENTRY 3 –
LEVEL 2

Health & Social Care
Workbook

Deborah Holder
Veronica Thomas

OXFORD
UNIVERSITY PRESS

OXFORD
UNIVERSITY PRESS

Great Clarendon Street, Oxford, OX2 6DP, United Kingdom

Oxford University Press is a department of the University of Oxford.
It furthers the University's objective of excellence in research, scholarship,
and education by publishing worldwide. Oxford is a registered trade mark of
Oxford University Press in the UK and in certain other countries

First published by Nelson Thornes Ltd in 2012
This edition published by Oxford University Press in 2014

British Library Cataloguing in Publication Data
Data available

978-1-4085-1833-5

10 9 8 7 6 5 4 3

Printed by Bell and Bain Ltd., Glasgow

MIX
Paper from
responsible sources
FSC
www.fsc.org FSC® C007785

Acknowledgements

Cover image: Kudryashka/Shutterstock
Illustrations: Pantek Media and Oxford Designers and Illustrators Ltd
Page make-up: Pantek Media, Maidstone

The author and the publisher would also like to thank the following for permission to reproduce
material:

Text
p10 adapted from Disabled Persons' Bill 1986; pp14, 112 Statistics on Alcohol: England, 2011.
Copyright © 2011, The Health and Social Care Information Centre. All rights reserved; p18 adapted
from 'What is a decibel and how it is measured?' © 1998–2012 HowStuffWorks, Inc; p18 from www.
hear.it.org; pp18, 126 from actiononhearingloss.org.uk; p22 Copyright © United States Food & Drug
Administration; p22 from 'Heart attack grill lives up to its name as customer collapses' 24 April
2012 © Telegraph Media Group Limited 2012; p26 Copyright BAAF and its suppliers © 1999–2012;
pp29, 109, 115 Office of National Statistics © Crown Copyright; pp29, 109 from 'Number of Future
Centenarians by Age Group' by Department for Work and Pension, April 2011 (c) Crown Copyright
2011; p33 adapted from 'The ageing population' © Parliamentary Copyright; pp73, 112 Reproduced
by kind permission of the Department of Health, © 2012; p106 Statistics on Obesity: England 2011.
Copyright © 2011, The Health and Social Care Information Centre. All rights reserved; Statistics on
Hearing Loss: England 2010. Copyright © 2010, The Health and Social

Care Information Centre. All rights reserved; pp133, 134 from FirstStop Care Advice © 2012 Elderly
Accomodation Counsel; p133 from 'NHS Personal Social Services – Gross Expenditure. Copyright ©
2011, The Health and Social Care Information Centre. All rights reserved.

All Crown Copyright material is reproduced under PSI licence no. C2009002012

Images
Alamy: p10cl (David Cole), p11 (Tim Gander), p33 (67photo), p69br (Paul Doyle), p85 (Andrew
Maynard), p89 (Cultura Creative), p93 (Corbis Bridge), p97 (Jack Sullivan), p119 (Ruby), p123 (Homer
Sykes); © Barnados 2012: p26, p114; Getty Images: p17 (Studio Box), p130 (Peter Cade); Mencap:
p10cr; Press Association Images: p13 (Tony Marshall/Empics Sport); Photofusion: p29 (Ulrike Preuss),
p69tr, p133 (Paula Solloway); Reuters Picture Library: p24 (Joshua Lott); Rex Features: p14 (Jason Bye);
Scope: p10bcl.

Although we have made every effort to trace and contact all
copyright holders before publication this has not been possible in all
cases. If notified, the publisher will rectify any errors or omissions at
the earliest opportunity.

Links to third party websites are provided by Oxford in good faith
and for information only. Oxford disclaims any responsibility for
the materials contained in any third party website referenced in
this work.

Contents

Introduction

'National Numeracy is a new charity focussed on transforming public attitudes to maths and numeracy and to seeing a measurable transformation of maths in school and for adults.

National Numeracy
for everyone, for life

We endorse the publisher's commitment to presenting mathematics in a range of adult-based contexts – school textbooks do not do this well. We also like the fact that they have used data from a range of sources and presented data in a variety of forms found in real-life examples.

To this end, these could be useful resources for encouraging renewed interest from adults in mathematics – mathematics that will appear more appropriate to their daily lives.'

Mike Ellicock
Chief Executive, National Numeracy
www.nationalnumeracy.org.uk

'Functional skills are the fundamental, applied skills in English, mathematics, and information and communication technology (ICT) which help people to gain the most from life, learning and work.'

Ofqual (2012), Criteria for Functional Skills Qualifications

This workbook is designed to present functional maths in a variety of contexts to make it accessible and relevant to you, as Health and Social Care candidates. It is intended to be written in, so use the extra white space for your workings out!

Being 'functional' means that you will:

- be able to apply skills to all sorts of real-life contexts
- have the mental ability to take on challenges in a range of new settings
- be able to work independently
- realise that tasks often need persistence, thought and reflection.

Features of this workbook are:

 FOCUS ON

Each Focus on is typically 1–2 pages long and will teach you specific Functional Skills. They include:

- guidance on the skill
- one or more activities to practise the skill.

 **SOURCE**

These pages will cover important aspects of Health and Social Care and consist of interesting source materials, such as newspaper articles or industry-related information, followed by various questions and activities for you to complete.

Good luck!

Number

Working with whole numbers

 FOCUS ON Number placement and calculations

Reading and writing numbers

It is important to recognise the value of digits in different columns so that you know the size of the number and can check that your answers are sensible.

	Millions			Thousands					
	H	T	U	H	T	U	H	T	U
Four thousand and sixty-five						4	0	6	5
Forty thousand and sixty-five					4	0	0	6	5
Four hundred thousand, six hundred and fifty				4	0	0	6	5	0
Four million, six hundred and fifty thousand			4	6	5	0	0	0	0
Forty million		4	0	0	0	0	0	0	0

HTU stands for hundreds, tens and units.

Rounding numbers

Rounded numbers can be used for estimation and checking. Numbers can be rounded to the nearest 10, 100, 1000, 10 000, etc. depending on the size of the number and the level of accuracy needed.

- To round to the nearest 100, for example, first find the digit in the hundreds column.

- If the 'deciding digit' in the column to the right is below 5, the digit in the hundreds column stays the same, i.e. the number is rounded down.

- If the 'deciding digit' in the column to the right is 5 or above, the digit in the hundreds column increases by 1, i.e. the number is rounded up.

Remember: if rounding to the nearest <u>10</u>, the number will end in <u>0</u>; to the nearest <u>100</u>, the number will end in <u>00</u>; to the nearest <u>1000</u> the number will end in <u>000</u> etc.

Examples:

Round 632 to the nearest 100 = 600

100 column   Below 5   *Rounds down*

Round 467 to the nearest 100 = 500

100 column   Above 5   *Rounds up*

Round 13 278 to the nearest 1000 = 13 000

1000 column   Below 5   *Rounds down*

Round 745 000 to the nearest 10 000 = 750 000

10 000 column   Equals 5   *Rounds up*

Calculating with whole numbers

Consider the problem and write it down in your own words if this helps.

Decide which numbers you need and whether you need to add, subtract, multiply, divide or carry out more than one of these operations in a particular order.

Check your answer is sensible based on your own knowledge and experience. You can also check it using the techniques shown below.

Checking calculations

Use estimation or the reverse operation to check your calculations.

Examples:

A charity raises £185 with the first fundraising event and £4212 with the second. How much more do they raise with the second event than the first? 4212 − 185 = 4027 Estimation check: 4200 − 200 = 4000, so the above answer is the right size. Reverse calculation check: 4027 + 185 = 4212	A social worker is paid £29 820 a year. How much is this a month? £29 820 ÷ 12 = £2485 Estimation check: 30 000 ÷ 10 = 3000, so the above answer is the right size. Reverse calculation check: £2485 × 12 = £29 820

Using formulae

When working with formulae, remember the order of operations can be represented by BIDMAS: **Brackets**, then **Indices**, **Division and Multiplication**, **Addition and Subtraction**.

Examples:

Formula for calculating cooking time of a turkey
$T = 28 (W + S) + 108$
Where T = time in minutes, W = weight of the turkey in kg and S = weight of stuffing in kg
If W = 4 and S = 0.25
4 + 0.25 = 4.25 $28 × 4.25 = 119$ $119 + 108 = 227$ minutes
Multiply as 28 is <u>next to</u> the bracket

Formula for calculating Body Mass Index (BMI)
$BMI = \dfrac{703 W}{h^2}$
Where W = weight in pounds and h = height in inches
If h = 72 and W = 168
$72 × 72 = 5184$ $703 × 168 = 118 104$ $118 104 ÷ 5184 = 22.8$
^{2} means <u>multiply number by itself</u>   *Divide as h^2 is <u>under</u> the line*

FOCUS ON Ratio

Writing ratios

Ratios can be expressed in different ways.

Examples:

'Only 1 in 4 of the staff were qualified' can be written as:
the ratio of qualified to unqualified staff was 1:4.

'Use 1 part disinfectant to 5 parts water' can be written as:
the ratio of disinfectant to water is 1:5.

Simplifying ratios

Simplify ratios by dividing both numbers by a common factor.

Examples:

5:10 can be simplified as 1:2 (divide each side by 5).
120:180 can be simplified as 2:3 (divide each side by 60).
Simplified ratios can then be used to calculate other quantities.

Calculating quantities using ratios

You can use a simple diagram to check that the numbers increase (or decrease) in proportion both horizontally and vertically.

Example:

If the ratio of staff to patients should be 1:4, how many patients can 12 staff care for?

Staff   Patients
   1 : 4   ×12
  12 : 48
      ×4
  12 × 4 = 48

You can also draw a diagram to help you visualise the problem.

Example:

1500 ml of diluted disinfectant is needed in a ratio of 2 parts disinfectant to 3 parts water. How much disinfectant and how much water is needed?

Divide the total amount by the total number of parts to find the value of 1 part. Use this to calculate quantities for 2 and 3 parts.

Check: 600 + 900 = 1500 ml

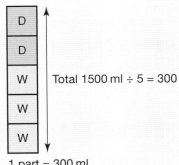

1 part = 300 ml
2 parts disinfectant = 600 ml
3 parts water = 900 ml

Working with fractions, decimals and percentages

 FOCUS ON Fractions, decimals and percentages

Fractions

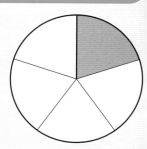

Fractions look like this and are made up of two different components.

$\frac{1}{5}$ numerator (number of pieces)
denominator (number of equal pieces the whole has been split into)

Adding and subtracting fractions

To add or subtract fractions, make the denominators the same. You may need to find an equivalent fraction. For example:

$\frac{1}{2} + \frac{2}{6}$

$\frac{1}{2} = \frac{3}{6}$ multiplying the top and bottom numbers by the same value, 3

so $\frac{3}{6} + \frac{2}{6} = \frac{5}{6}$

Or you can multiply the denominators together to make them the same.

For example: For $\frac{1}{2} + \frac{2}{6}$, 2 × 6 = 12

Then cross-multiply the top and bottom numbers to make equal fractions:

so $\frac{1}{2} + \frac{2}{6} = \frac{1 \times 6 + 2 \times 2}{12}$

$= \frac{10}{12}$

Then cancel to the lowest term.

$\frac{10}{12} = \frac{5}{6}$

Multiplying fractions

Multiply the numerators to give the new numerator. Multiply the denominators to give the new denominator. For example:

$\frac{1}{2} \times \frac{1}{4} = \frac{1}{8}$

In words, the above example could be thought of as half of a quarter.

Dividing fractions

Turn one fraction upside-down and then multiply the numerators and denominators as above. For example:

$\frac{1}{2} \div \frac{1}{4} = \frac{1}{2} \times \frac{4}{1} = \frac{4}{2} = 2$

Percentages

'Percent' means a specific proportion of every 100.
For example:

1% = 1p in every £   or   £1 in every £100
10% = 10 in every 100 = $\frac{10}{100} = \frac{1}{10}$

> **Remember:** there are 100 pennies in every pound.

To find $\frac{1}{10}$ of a number divide it by 10. Move the numbers one column to the right around the decimal point. For example: 10% of 45 m = 4.5 m

With percentages, always remember to calculate from the starting point. For example, if a shop reduces a £10 game by 10% and then later decides to increase the price by 10%, the price will not return to £10.

Start with £10. 10% of £10 is £1, so the reduced price is £10 - £1 = £9.
Start with £9. 10% of £9 is 90p, so the increased price is £9 + 90p = £9.90.

Decimals

Remember the place values in decimals. This is how decimals are set out:

hundreds	tens	units	decimal point	tenths	hundredths	thousandths	2 tenths of whatever the unit is
		0	.	2			

Example: 0.2 m = $\frac{2}{10}$ of a metre = 20 cm

Always write a zero before the decimal point. This avoids any confusion with the decimal point.

Remember: there are 100 centimetres in every metre.

Zeros after the final number following a decimal point are not needed. For example: 0.2 = 0.20 = 0.200. However, remember that you always need two decimal places for money.

Adding and subtracting decimals

All the numbers must be in the same units, for example, all centimetres or all metres. Convert them if not. Set out the sum so that all the decimal points align vertically and then add or subtract the numbers in columns.

Multiplying decimals

Ignore the decimal point to start with. Multiply the numbers normally. When you have the numbers in the answer, count the total number of digits after all the decimal points in the sum. Insert the decimal point so that the same number of digits are to the right of it in the answer. For example:

$$\begin{array}{r} 3.14 \\ \times\ 1.6 \\ \hline 5.024 \end{array}$$

3 digits to the right of all the points

3 digits to the right of the point

Dividing decimals

Multiply both numbers by 10, 100, etc. to remove the decimal point from the dividing number. Then divide as normal. For example:

$$3.5\overline{)70.7} \qquad\qquad 35\overline{)707.0} = 20.2$$

Recurring decimals

If a fraction does not divide exactly to make an equivalent decimal value, but has a recurring number or numbers, it is called a recurring decimal. A dot is written above the recurring numbers. For example $\frac{1}{3}$ = 0.333... = 0.$\dot{3}$

Working with whole numbers

 **Changing attitudes to disability**

Key dates
The following dates are key landmarks in changing attitudes to disability.

1986 Disabled Persons Bill
Gave disabled people the right to have their needs assessed and met.

2005 Government Report 'Improving the Life Chances of Disabled People' published. Set out plans to improve the quality of life of disabled adults and children by 2025.

1944 Disabled Persons (Employment) Act
Required employers with 20 or more staff to ensure that 3% of employees were registered disabled.

1913 Mental Deficiency Act
Children identified as 'defective' were sent to live in institutions.

1970 Chronically Sick and Disabled Persons Act
Required local authorities to improve services to disabled people.

1995 Disability Discrimination Act (DDA)
Made discrimination against disabled people illegal.

1886 Idiots Act
First legislation dealing with the educational needs of those with learning disability. It made the distinction between lunatics and idiots and imbeciles.

1991 Introduction of Disability Living Allowance and Disability Working Allowance

1981 Education Act
Laid down that children with 'special needs' should be educated in mainstream schools or classes where possible.

1952 Formation of the Spastics Society
Set up to campaign for better education and other opportunities for people with cerebral palsy. This later became Scope.

1946 Formation of the National Association of Parents of Backward Children
This later became Mencap.

1927 Mental Deficiency (Amendment) Act
Replaced term 'moral defective' with 'moral imbecile'. Allowed for mental deficiency resulting from illness.

2001 Special Education Needs and Disability Act
Extended disability discrimination legislation to schools.

1845 Lunacy Act
Made no distinction between learning disability and mental illness. Stated that 'lunatic shall mean insane person or any person being idiot or lunatic or of unsound mind'.

E3

1 List the key landmarks in changing attitudes to disability in date order, starting with the oldest and ending with the most recent. Write down the year and the title or description given in bold.

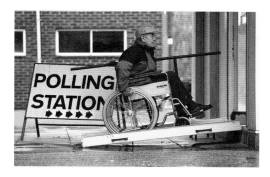

2 How many years were there between:

a) the Chronically Sick and Disabled Persons Act and the Education Act?

b) the Disabled Persons Bill and the introduction of Disability Living Allowance and Disability Working Allowance?

c) the Disability Discrimination Act and the Special Education Needs and Disability Act?

3 How many years ago was the government report 'Improving the Life Chances of Disabled People' published?

L1

1 How many years ago was the Lunacy Act introduced?

[ ]

2 How many years were there between the Lunacy Act and the Idiots Act?

[ ]

3 What was introduced 100 years after the Idiots Act?

[ ]

4 How many years after the formation of the National Association of Parents of Backward Children was the Spastics Society formed?

[ ]

5 How old would someone who is 25 now have been when the Special Education Needs and Disability Act was introduced?

[ ]

6 How old would someone who is 50 now have been when the Disability Discrimination Act was introduced?

[ ]

7 How old would someone be now who was 18 when the Disability Working Allowance was introduced?

[ ]

8 How old would someone be now who was 18 in the year the Disabled Persons (Employment) Act was passed?

[ ]

9 How old would someone be now who was born in the year the Spastics Society was formed?

[ ]

10 How old would someone be now who was born in the year the National Association of Parents of Backward Children was formed?

[ ]

L2

1   Draw a timeline down the page with a suitable scale. Mark the key landmarks in changing attitudes to disability in date order on the timeline.

2   Add the following events to your timeline:

a) 1948 first International Wheelchair Games

b) 1960 first Paralympic Summer Games

c) 1976 first Paralympic Winter Games.

Timeline

Working with whole numbers

 **Alcohol and health**

Records show that the number of alcohol-related deaths and illnesses is increasing. A health organisation wants to use some of the available data to warn the public about the dangers of alcohol.

Alcohol-related deaths 2001–2009									
Year	2001	2002	2003	2004	2005	2006	2007	2008	2009
Men	3576	3631	3970	3922	4096	4272	4236	4473	4316
Women	1900	1951	2011	2114	2095	2245	2305	2295	2268

Sample of alcohol-related NHS hospital admissions 2009/10		
Cause	Number of males	Number of females
Fall injuries	14 400	9 400
Work- or machine-related	1 200	300
Drowning	500	200
Accidental excessive cold	100	100
Intentional self-harm	7 700	11 900
Assault	7 100	1 400
Pedestrian traffic accident	900	200
Road traffic accident	3 500	700

Number of alcohol-related NHS hospital admissions 2002/03–2009/10							
Year	2002/03	2003/04	2004/05	2005/06	2006/07	2007/08	2008/09	2009/10
Number	510 800	570 100	644 700	736 100	802 100	863 600	945 500	1 057 000

E3

Study the table of alcohol-related deaths and answer the following questions.

1 Which year had the greatest number of alcohol-related deaths for:

a) men?

b) women?

2 Which year had the lowest number of alcohol-related deaths for:

a) men?

b) women?

3 Complete the table, rounding the numbers to the nearest 100. Use your rounded figures to estimate the total number of deaths for each year.

Number of alcohol-related deaths (to the nearest 100)									
Year	2001	2002	2003	2004	2005	2006	2007	2008	2009
Men									
Women									
Total									

4 The health organisation want to produce a leaflet to warn people about the dangers of drinking.

Use your rounded figures to fill in the missing numbers.

a) Between 2001 and 2009, the number of alcohol-related deaths for men rose by

.

b) Between 2001 and 2009, the number of alcohol-related deaths for women rose by

.

c) Between 2001 and 2009 the total number of alcohol-related deaths rose by

.

5 What is the actual total number of alcohol-related deaths for:

a) 2001?

b) 2009?

> Use the figures from the table on the source sheet.

6 What is the actual increase in the total number of alcohol-related deaths between 2001 and 2009?

7 Use your answer to Question 6. What is the difference between the actual figure and the rounded figure?

L1

Study the table of alcohol-related hospital admissions in 2009/10 and answer the following questions.

1 What is the difference between the number of admissions for 'fall injuries' for males and females?

2 How many more males than females were admitted for 'road traffic accidents'?

3 How many more females than males were admitted for 'intentional self-harm'?

4 a) Complete the table below with the number of admissions for males and females for each cause and then calculate the grand total for all admissions.

   b) Round each of the admissions numbers to the nearest thousand and then find the estimated total.

Cause	Total (male and female)	Total rounded to nearest 1000
Fall injuries		
Work- or machine-related		
Drowning		
Accidental excessive cold		
Intentional self-harm		
Assault		
Pedestrian traffic accident		
Road traffic accident		
	Grand total:	Estimated total:

5 What is the difference between the actual and estimated totals?

6 Write a suitable statement highlighting the number of people who suffer fall injuries due to alcohol.

L2

Study the table of alcohol-related NHS hospital admissions 2002/03–2009/10 and answer the following questions.

1 Complete the table below, rounding the actual numbers to the nearest 1000, 10 000 and 100 000.

Year	2002/03	2003/04	2004/05	2005/06	2006/07	2007/08	2008/09	2009/10
Number	510 800	570 100	644 700	736 100	802 100	863 600	945 500	1 057 000
Rounded to nearest 1000								
Rounded to nearest 10 000								
Rounded to nearest 100 000								

2 a) How many more actual admissions were there in 2009/10 than in 2002/03?

b) Show how you can use your rounded figures to check your answer.

3 Write a headline highlighting the number of people who were admitted to hospital for alcohol-related illnesses.

Working with whole numbers

SOURCE Assessing hearing loss

Kamisha is a trainee care worker who is going to work with adults who have suffered hearing loss. She wants to have a greater understanding of how this disability affects their lives, and has gathered the following information.

Measuring hearing

The decibel (dB) is the unit used to measure the intensity of a sound.

On the decibel scale, the smallest sound you can hear is 0 dB.

* A sound 10 times more powerful than the smallest sound is 10 dB.
* A sound 100 times more powerful than the smallest sound is 20 dB.
* A sound 1000 times more powerful than the smallest sound is 30 dB.

Sounds and their ratings in decibels

A Normal breathing 10	**F** Lawnmower 90	**K** Normal conversation 60
B Car horn 110	**G** Rainfall 40	**L** Watch ticking 30
C Passing car 80	**H** Ambulance siren 120	**M** Motorbike 90
D Laughter 65	**I** Fireworks 145	**N** Vacuum cleaner 70
E Jet plane 135	**J** Washing machine 75	**O** Mosquito 20

Definitions of hearing loss

* Mild Hearing Loss is defined as when the quietest sounds that can be heard are between 25 and 39 dB.

* Moderate Hearing Loss is defined as when the quietest sounds that can be heard are between 40 and 69 dB.

* Severe Hearing Loss is defined as when the quietest sounds that can be heard are between 70 and 94 dB.

* Profound Hearing Loss is defined as when the quietest sounds that can be heard are 95 dB and above.

Number of people with hearing loss by age in the UK (2010)		
Age band	**Total number with hearing loss**	**Number with severe/profound hearing loss**
16–49	1 157 500	36 000
50–64	2 563 500	99 500
65–79	3 768 000	211 000
80+	2 622 500	474 500

E3

1 Using the definitions of hearing loss on the source sheet, complete the table by deciding which of the sounds in the data can be heard, given their decibel rating. Write each letter in the correct box.

Hearing loss	Cannot hear	May be able to hear	Should be able to hear
Mild			
Moderate			
Severe			
Profound			

2 Do you think someone with mild, moderate, severe or profound hearing loss would be able to hear the following? Give reasons for your answers.

a) A television

b) A doorbell

c) A fire alarm

L1

1 Decide whether the following statements are true or false.

a) Over $2\frac{1}{2}$ million people aged 50–64 have hearing loss.

True ☐   False ☐

b) Just under ten thousand people aged 50–64 have severe/profound hearing loss.

True ☐   False ☐

c) Just under half a million people aged 80+ have severe/profound hearing loss.

True ☐   False ☐

d) The number of people with hearing loss aged 50–64 is more than double the number aged 16–49.

True ☐   False ☐

e) The number of people with severe/profound hearing loss aged 65–79 is more than ten times the number aged 50–64.

True ☐   False ☐

f) The total number of people with hearing loss is over ten million.

True ☐   False ☐

g) More than three-quarters of a million people have severe/profound hearing loss.

True ☐   False ☐

2 Calculate the total number of people aged 16–80+ with severe/profound hearing loss. Give your answer in:

a) figures

b) words

3 How many more people aged 80+ than people aged 16–49 had severe/profound hearing loss? Give your answer in:

a) figures

b) words

4 Calculate the total number of people aged 16–80+ with hearing loss.

Give your answer in:

a) figures

b) words

5 How many more people aged 80+ than people aged 16–49 had hearing loss?

Give your answer in:

a) figures

b) words

L2

1 Read how the decibel scale works and look at the examples.

How many times more powerful than the smallest sound you can hear is:

a) normal breathing?

b) a mosquito?

c) rainfall?

d) normal conversation?

e) a vacuum cleaner?

f) a passing car?

g) a lawnmower?

h) a car horn?

i) an ambulance siren?

2 Use your answers to question 1 to decide whether the following statements are true or false.

a) The sound of a watch ticking is 100 times more powerful than normal breathing.

True ☐   False ☐

b) The sound of normal conversation is twice as powerful as that of rainfall.

True ☐   False ☐

c) Normal conversation is a million times more powerful than the smallest sound you can hear.

True ☐   False ☐

d) The sound of a passing car is 1000 times more powerful than that of normal conversation.

True ☐   False ☐

e) The sound of a passing car has one-tenth of the power that the sound of a lawnmower has.

True ☐   False ☐

f) The sound of a lawnmower is a billion times more powerful than the smallest sound.

True ☐   False ☐

g) The sound of an ambulance siren is a trillion times more powerful than the smallest sound.

True ☐   False ☐

Working with whole numbers

 **SOURCE** Choosing a healthy diet

Sylvie is a community health worker who is supporting a group of young mothers. She wants them to understand the labelling on food packaging so that they can make healthy choices, and has collected information to use with them. Sylvie also wants to make them aware of the dangers of eating too much junk food and has found an extreme example to catch their attention!

Guideline daily amount (GDA)	Calories	Sugars	Fat	Saturates	Salt
Women	2000	90 g	70 g	20 g	6 g
Men	2500	120 g	95 g	30 g	6 g

Prepared food GDA labels

Per serving: **Chicken casserole**

Calories	Sugars	Fat	Saturates	Salt
315	1 g	19 g	6 g	2 g

Per serving: **Beef lasagne**

Calories	Sugars	Fat	Saturates	Salt
554	8 g	28 g	15 g	3 g

Per serving: **Chilli and rice**

Calories	Sugars	Fat	Saturates	Salt
515	3 g	10 g	3 g	3 g

Per serving: **Vegetable lasagne**

Calories	Sugars	Fat	Saturates	Salt
364	13 g	13 g	8 g	2 g

Per serving: **Chicken tikka and rice**

Calories	Sugars	Fat	Saturates	Salt
905	19 g	40 g	14 g	4 g

Per serving: **Macaroni cheese**

Calories	Sugars	Fat	Saturates	Salt
975	7 g	33 g	17 g	3 g

Per serving: **Cheese pizza**

Calories	Sugars	Fat	Saturates	Salt
334	4 g	14 g	7 g	1 g

Heart Attack Grill lives up to its name as customer collapses

A restaurant in Las Vegas that revels in serving oversized hamburgers has claimed its second victim this year after a customer collapsed with a suspected cardiac arrest at the weekend.

According to reports in the US, the unnamed woman was eating a Double Bypass sandwich at the Heart Attack Grill, a hospital-themed diner with the motto of 'Taste Worth Dying For'.

All the diner's burger buns are smothered in lard, while the Quadruple Bypass has **four half-pound meat patties** and **eight slices of processed American cheese**. A side order of **20 slices of bacon** is available for $3.69 (£2.50).

The restaurant has been awarded the Guinness World Record for 'Highest Calorie Hamburger' for its **9983 calorie** 'Quadruple Bypass Burger'.

The restaurant grants **free meals to customers who weigh more than 350 pounds** and prides itself on taking a stand against diet culture. Staff members dress in medical gear and an ambulance is parked outside.

E3

1 Which prepared food has the most calories?

2 Which prepared food has the most fat?

3 What is the difference between the number of calories in the beef lasagne and the vegetable lasagne?

4 Which prepared food would you recommend to someone on a low-calorie diet?

5 Which prepared food would you recommend to someone on a low-fat diet?

6 Which prepared food would you not recommend to someone on a low-salt diet?

7 What is the difference between the number of calories in a serving of cheese pizza and the GDA for:

a) women?

b) men?

8 What is the difference between the number of calories in a 'Quadruple Bypass' burger and the GDA for:

a) women?

b) men?

9 Round the number of calories in a 'Quadruple Bypass' to the nearest 1000.

10 a) Round the number of calories in a serving of macaroni cheese to the nearest 100.

   b) Use that figure to estimate how many servings of macaroni cheese would contain the same number of calories as a 'Quadruple Bypass'.

11 a) Round the number of calories in a serving of chilli and rice to the nearest 100.

   b) Use that figure to estimate how many servings would contain the same number of calories as a 'Quadruple Bypass'.

L1

1 Estimate how many times the GDA for:

a) women [ ] b) men [ ]

is contained in the number of calories in a 'Quadruple Bypass'.

2 Using the information in the table below, estimate the amount of fat (in grams) in a 'Quadruple Bypass' plus a side order of bacon. Show your workings.

Quadruple Bypass Burger®

$12.95

	Fat content	Salt content
Half-pound meat pattie	50 g	175 mg
Slice of processed cheese	4 g	50 mg
Slice of bacon	3 g	180 mg

3 Using your answer to question 2, comment on the amount of fat in a 'Quadruple Bypass' plus a side order of bacon, compared with the GDA for women and men.

4 Using the information above, estimate the amount of salt (in milligrams) in a 'Quadruple Bypass' plus a side order of bacon. Show your workings.

5 a) Convert your answer to question 4 into grams. **Remember:** 1000 mg = 1 g.

[ ]

b) Comment on the amount of salt in a 'Quadruple Bypass' plus a side order of bacon, compared with the GDA for women and men.

6 There are 14 pounds in a stone. What is the minimum weight in stones and pounds of Heart Attack Grill customers who get free meals?

[ ]

L2

1 Using the formula for calculating BMI and rounding your answers to the nearest whole number, calculate the BMI for someone who weighs 350 pounds and is:

a) 6 feet tall

Remember: the formula is

$$BMI = \frac{703W}{h^2} \text{ (W = weight in pounds h = height in inches).}$$

b) 5 feet 4 inches tall.

Remember: 1 ft = 12 in.

2 Referring to the table and giving your answers in stones and pounds to the nearest pound, calculate the maximum weight someone classed as 'normal' can be if they are:

a) 6 feet tall

BMI	Weight status
Below 18.5	Underweight
18.5–24.9	Normal
25–29.9	Overweight
30 and above	Obese

b) 5 feet 4 inches tall.

3 How tall would someone who weighed 350 pounds have to be if they were classed as 'overweight' rather than 'obese'? Show your workings.

Working with whole numbers

SOURCE Number of children in public care, 2011

A team of social workers is looking at how resources can best be used to support children in care. They are interested in the total number of children in care and the number adopted by age group and gender.

Adoption from care

- Four thousand, four hundred and seventy-two adoption orders were made in England and Wales during 2010.

- Three thousand and sixty children were adopted from care during the year ending 31 March 2011.

- One thousand, five hundred and sixty of these were boys.

- One thousand, five hundred were girls.

- Sixty children were under 1 year old.

- Two thousand, one hundred and seventy were aged between 1 and 4 years old.

- Seven hundred and thirty were aged between 5 and 9 years old.

- Ninety were aged between 10 and 15 years old.

Total number of children in care

- Sixty-five thousand, five hundred and twenty children were in the care of local authorities on 31 March 2011.

- Thirty-six thousand, four hundred and seventy of these were boys.

- Twenty-nine thousand and fifty were girls.

- Three thousand, six hundred and sixty were under 1 year old.

- Twelve thousand and twenty were aged between 1 and 4 years old.

- Eleven thousand, eight hundred and thirty were aged between 5 and 9 years old.

- Twenty-four thousand, one hundred and sixty were aged between 10 and 15 years old.

- Thirteen thousand, eight hundred and sixty were aged 16 and over.

'If it wasn't for my foster parents, Alan and Sarah, I wouldn't be the football coach I am today.'

It takes someone special to foster or adopt a child. You not only look after their practical needs but also give them valuable emotional support. That way they have a real chance to achieve their goals.

Barnardo's urgently needs more foster carers and adoptive parents.
To find out more visit us today at
www.barnardos.org.uk

Believe in children
Barnardo's

E3

Study the statistics on adoption and answer the following questions.

Rewrite in figures the numbers given in words.

a) [   ]   e) [   ]

b) [   ]   f) [   ]

c) [   ]   g) [   ]

d) [   ]   h) [   ]

Write your answers to the following questions in words.

a) How many more boys than girls were adopted?

[   ]

b) How many more children aged 1–4 were adopted than children aged 5–9?

[   ]

c) How many more children aged 10–15 were adopted than children under 1 year old?

[   ]

d) What was the total number of children adopted aged 0–4?

[   ]

e) What was the total number of children adopted aged 5–15?

[   ]

L1

Study the statistics on children in care and answer the following questions.

Rewrite in figures the numbers given in words.

a) [   ]   e) [   ]

b) [   ]   f) [   ]

c) [   ]   g) [   ]

d) [   ]   h) [   ]

2 Write the answers to the following questions in words.

a) What was the difference between the number of boys and girls in care?

b) How many of these children were under 5 years old?

c) How many of these children were aged 5–15 years?

d) What was the difference between the number aged 10–15 and the number over 16?

L2

Study the statistics on adoption and children in care, then answer the following questions. The number of adopted children is included in the number in care.

1 Use the information to calculate (i) the number of children that were not adopted and (ii) the approximate ratio of adopted to non-adopted children in care for:

a) children under 1 year old

i)

ii)

b) children aged 1–4

i)

ii)

c) children aged 5–9

i)

ii)

d) children aged 10–15

i)

ii)

Working with whole numbers

SOURCE The ageing population – centenarians

Dominika works for a national charity that gives advice and help to older people. She is interested in population trends for older people, as this will help the charity to plan areas of work for future development and fundraising campaigns to support this.

Estimates of centenarians in the UK

Over the last 30 years the number of centenarians (people aged 100 years or more) in the UK has increased five-fold from 2500 in 1980 to 12 640 in 2010.

The chart below shows the estimated number of centenarians in the UK for the period 1965 to 2010. Throughout this period female centenarians have always outnumbered male centenarians due to higher life expectancies for women. The number of female centenarians has risen steadily since the mid-1960s and this increase has accelerated over the last decade.

The number of male centenarians has also shown a marked increase since the year 2000.

The ratio of female to male centenarians has started to fall in recent years; in 2000 there were approximately nine female centenarians for every male centenarian, in 2009 there were approximately six female centenarians for every male centenarian and in 2010 the ratio fell to approximately five female centenarians for every male centenarian.

Estimated population aged 100 years and over, 1970–2010					
	1970	**1980**	**1990**	**2000**	**2010**
England and Wales	1080	2280	4030	6230	11 610
Scotland	80	150	260	480	820
Northern Ireland	30	70	100	140	210

Source: Office for National Statistics

Projected number of centenarians in the UK						
Year	**100+**			**110+**		
	Male	**Female**	**All**	**Male**	**Female**	**All**
2021	6500	19 400	25 900	0	0	0
2031	20 500	45 000	65 500	0	100	100
2041	54 200	100 700	154 900	100	300	400
2051	102 000	179 200	281 200	700	1600	2300
2061	148 300	250 300	398 600	1900	3900	5800
2071	219 000	356 500	575 500	3900	7800	11 700
2081	257 700	393 300	651 000	7700	14 300	22 000

E3

Study the chart on estimated population aged 100 or over, and answer these questions.

1   Calculate the totals for each year for the UK and add them to the table below.

	1970	1980	1990	2000	2010
UK total					

2   Round the figures for 2010 to the nearest 100. Then use the rounded figures to check your answer above.

	Rounded figure (to nearest 100)
England and Wales	
Scotland	
Northern Ireland	
Estimated total	

3   Decide whether the following statements are true or false.

a) There were 10 530 more centenarians in 2010 in England and Wales than in 1970.

   True ☐   False ☐

b) The number of centenarians in Scotland increased by 460 between 2000 and 2010.

   True ☐   False ☐

c) There were three times as many centenarians in Northern Ireland in 2010 as there were in 1980.

   True ☐   False ☐

d) There were 610 more centenarians in Scotland than in Northern Ireland in 2010.

   True ☐   False ☐

e) The number of centenarians in England and Wales doubled between 2000 and 2010.

   True ☐   False ☐

f) There were more than twelve thousand five hundred centenarians in the UK in 2010.

   True ☐   False ☐

L1

Study the chart on estimated population aged 100 or over, and answer these questions.

1 In 2000 the ratio of male to female centenarians was approximately 1:9.

Use this ratio to estimate how many male and female centenarians there were in 2000 in:

a) England and Wales

b) Scotland

c) Northern Ireland

2 In 2010 the ratio of male to female centenarians was approximately 1:5.

Use this ratio to estimate how many male and female centenarians there were in 2010 in:

a) England and Wales

b) Scotland

c) Northern Ireland

Study the chart projecting the number of centenarians in the UK and answer the following questions.

3 In what year does the projected number of people aged 100 years and above reach:

a) more than a quarter of a million?

b) more than half a million?

4 What is the difference between the projected number of people aged 100 years and above in 2021 and 2081?

L2

Study the chart projecting the number of centenarians in the UK and answer the following questions.

1 What is the projected ratio, in its simplest form, of men to women who are 110+ for:

a) 2041?

b) 2081?

2 Round the figures appropriately and calculate the projected ratio of men to women who are 100+ for:

a) 2021

b) 2031

c) 2041

d) 2051

e) 2061

f) 2071

g) 2081

Working with whole numbers

 **SOURCE** Ageing population – care requirements

Philip is a student nurse currently on placement on a ward for older people. He is interested in the following articles as he is thinking of specialising in this area of work and wants to know how good the career prospects are. Some of the information is also useful to him in evaluating his placement.

Caring for older people

The UK's ageing population has considerable consequences for public services.

Ten million people in the UK are over 65 years old. The latest projections are for 5½ million more elderly people in 20 years' time, and the number will have nearly doubled to around 19 million by 2050.

While one in six of the UK population is currently aged 65 and over, by 2050 one in four will be.

In 2008 there were 3.2 people of working age for every person of pensionable age. This ratio is projected to fall to 2.8 by 2033.

Nurse-to-patient ratios must increase to improve safety

The Royal College of Nursing has issued unprecedented guidance to hospital managers on staffing levels and skill mix for older people's wards.

Nurse-to-patient ratios on older people's wards		
	Nurse : patient ratio	Staff* : patient ratio
Current	1 : 9	1 : 4.6
Basic safe care	1 : 7	1 : 3.3–1 : 3.8
Ideal, good-quality care	1 : 5–1 : 7	1 : 3.3–1 : 3.8

* Staff are in addition to nurses.

E3

Study the report on care of the elderly and answer the following questions.

1 One in six people aged 65 or over can be represented in the following way:

Aged 65 and over   Under 65

Draw a similar diagram to represent one in four people aged 65 or over.

2 If one in 6 people are currently aged 65 or over:

a) how many people out of 12 will be?

b) how many people out of 18 will be?

c) how many people out of 60 will be?

d) how many people out of 600 will be?

3 If one in 4 people will be aged 65 or over in 2050:

a) how many people out of 8 will be?

b) how many people out of 12 will be?

c) how many people out of 24 will be?

d) how many people out of 240 will be?

4 What fraction of the population are currently aged 65 or over?

5 What fraction of the population will be aged 65 or over in 2050?

L1

1 Write 'one in 6' as a ratio.

2 Write 'one in 4' as a ratio.

Study the article on nurse-to-patient ratios and answer the following questions.

3 How many patients can 2 nurses oversee in order to satisfy:

a) current standards?

b) basic safe care?

c) ideal good-quality care?

4 How many patients can 6 nurses oversee in order to satisfy:

a) current standards?

b) basic safe care?

c) ideal good-quality care?

5 If there are 35 patients, what is the minimum number of nurses needed for the following situations? Give reasons for your answers.

a) Current standards

b) Basic safe care

c) Ideal good-quality care

6 What is the minimum number of staff needed for 9 patients for the following situations? Show your workings and give reasons for your answers.

a) Current standards

b) Basic safe care

L2

Study the article on nurse-to-patient ratios and answer the following questions.

1   How many patients can 3 nurses oversee in the following situations?

   a) Current standards

   b) Basic safe care

2   How many staff would be needed in each case? Show your workings and give reasons for your answers.

   a)

   b)

3   How many patients can 5 nurses oversee in the following situations?

   a) Current standards

   b) Basic safe care

4   How many staff would be needed in each case? Show your workings and give reasons for your answers.

   a)

   b)

5   What is the minimum number of nurses and staff needed for 36 patients in order to satisfy:

   a) current standards?

   b) basic safe care?

6   What is the minimum number of nurses and staff needed for 60 patients in order to satisfy:

   a) current standards?

   b) basic safe care?

Working with fractions, percentages and decimals

SOURCE Catering for eating disorders

When young people arrive at the Eating Disorder Unit they sometimes start on quarter food portions. By day 3 they are expected to eat half portions, and full portions by the end of the week.

Examples of full food portions

Full portion =
1 breakfast biscuit

Full portion = 200 ml

Full portion = all 4 pieces
(2 slices of bread)

Full portion =
the whole pizza

Full portion = $\frac{1}{3}$ of
the meat pie

Full portion of peas, beans
or sweetcorn = 100 g

Full portion = 1 potato

Full portion = $\frac{1}{4}$ fruit
crumble

Full portion = 1 apple

E3

1 Draw lines to match the fraction of pizza required for $\frac{1}{4}$ and $\frac{1}{2}$ portions.

$\frac{2}{8}$

$\frac{1}{4}$

$\frac{1}{8}$

$\frac{4}{8}$

$\frac{1}{2}$

$\frac{3}{8}$

2 How much bread is needed to make:

a) a full sandwich?

b) half a sandwich?

c) quarter of a sandwich?

3 What weight of beans, peas or sweetcorn is needed for:

a) a full portion?

b) half a portion?

c) quarter of a portion?

4 How much milk is needed for:

a) a full portion?

b) half a portion?

c) quarter of a portion?

5 Eight young people in the unit are going to have a meal of pizza, potato and beans. Three are on $\frac{1}{4}$ portions, 3 are on $\frac{1}{2}$ portions and 2 are on full portions.

a) How many potatoes need to be cooked?

b) What weight of beans should be cooked?

c) How many pizzas need to be cooked?

d) If they all eat what they should, what fraction of a pizza will be left?

L1

There are 8 young people in the unit. Three are on $\frac{1}{4}$ portions, 3 are on $\frac{1}{2}$ portions and the rest are on full portions. Use the information on full food portions to answer the following questions.

1 How many biscuits are needed for breakfast?

2 A new one-litre bottle of milk is opened.

   a) How much milk is needed?

   b) Write as a fraction the amount of milk left in the bottle.

3 There are 30 slices of bread in a loaf.

   a) How many slices of bread are used to make the sandwiches?

   b) Write as a fraction the number of slices left in the loaf.

4 What fraction of the meat pie is needed for:

   a) a full portion?

   b) half a portion?

   c) quarter of a portion?

   d) How many pies need to be cooked for the whole group?

   e) What fraction of a pie will be left over?

5 What fraction of a potato will be left over?

6 A bag of sweetcorn weighs 500g. What fraction of the bag will be left over?

7 If the young people all have apples, what fraction of an apple is left over?

L2

The chart on the right shows the numbers of young people on the different food portions over the next 3 days. Use the information on full food portions to answer the following questions.

	Monday	Tuesday	Wednesday
$\frac{1}{4}$ portions	3	2	1
$\frac{1}{2}$ portions	3	2	3
Full portions	2	4	4

1   There are 14 breakfast biscuits left in the pack. Joel says this is enough for all breakfasts on Monday, Tuesday and Wednesday. Is he right? Explain.

2   A loaf of bread has 30 slices. Approximately what fraction of the loaf will be left after making sandwiches on Tuesday?

3   What fraction of the fruit crumble is needed for:

a) a full portion?

b) half a portion?

c) quarter of a portion?

d) How many fruit crumbles need to be cooked for the group on Monday?

e) What fraction of a fruit crumble will be left over?

4   Emma is on half portions, but is having a bad day. She only eats half of what she should. Write what she does eat as a fraction of:

a) a full sandwich portion

b) a meat pie

c) a fruit crumble.

5   Fatima is on half portions, but only eats a quarter of what she should. Write what she does eat as a fraction of:

a) a full sandwich portion

b) a fruit crumble.

Working with fractions, percentages and decimals

Choosing takeaway meals

Adults in supported living want to have their favourite takeaway meals for a special night in, so support workers have collected these takeaway menus.

Menu 1

Chinese House

Set menu for 2   £18

10% off for collection in addition to other offers

Special Offer 12% off set menu

Prawn crackers
Chicken soup
Sweet and sour chicken
Yung chow fried rice

Main menu

Barbecued spare ribs	£4.35
Sweet and sour pork	£4.20
Szechuan beef	£4.20
Crispy chicken with beansprouts	£4.50
Beef chop suey	£3.90
Special fried rice	£3.60
Steamed or fried rice	£1.80

$\frac{1}{3}$ off main orders before 6:30pm

Menu 2

Balti Corner

Balti specialities

10% off for collection

	Meat	Prawn	Veg
Balti	£4.90	£7.30	£4.40
Dhansak	£5.30	£7.20	£4.30
Dopiaza	£5.20	£7.40	£4.50
Biryani	£6.50	£7.90	£4.70

Popular curry dishes

15% off for collection

	Meat	Prawn	Veg
Curry	£4.80	£5.60	£4.60
Jalfrezi	£5.20	£5.70	£4.30
Tikka	£5.90	£6.10	£4.20

Fried or pilau rice £1.99
Nan bread £1.50

Menu 3

Burger and Pizza House

20% off meal deals

5% off all pizzas

15% of all side orders

Mega Meal Deal

Large, med & small pizza
Potato wedges, coleslaw, salad, fries, 2 dips, drink   **£19.95**

Family Meal Deal

Large, med pizza
10 chicken wings, salad, fries, garlic bread, 2 dips, drink **£14.95**

Bumper Burger Deal

4 burgers, wedges, sweetcorn, dips, drink   **£12.50**

Side orders

Wedges	**£1.50**
Coleslaw	**90p**
Sweetcorn	**95p**

Pizzas

Large pizza + 3 toppings	**£8.99**
Med pizza + 3 toppings	**£5.99**
Small pizza + 3 toppings	**£4.50**

E3

1 The residents would like to order the following from Chinese House:

   They order before 6:30pm but are not going to collect. Find the discount, in money, and the cost, with discount, for each item.

   1 × beef chop suey
   1 × crispy chicken with beansprouts
   1 × Szechuan beef
   1 × special fried rice
   1 × fried rice

   a) Beef chop suey

   Discount [   ]   Cost [   ]

   b) Crispy chicken with beansprouts

   Discount [   ]   Cost [   ]

   c) Szechuan beef

   Discount [   ]   Cost [   ]

   d) Special fried rice

   Discount [   ]   Cost [   ]

   e) Fried rice

   Discount [   ]   Cost [   ]

   f) What will the food cost in total?

   [   ]

2 Tim finds a different discount from Chinese House. You can buy two meals and get the cheapest one at half price. All rice dishes have a quarter off the price. These discounts cannot be used with any other offers.

   a) Which meal will be half price? [   ]

   b) What will it cost with this discount on the full price? [   ]

   c) What will the two rice dishes cost? [   ]

   d) What will the food cost in total? [   ]

3 Which discount would you recommend? Explain why.

L1

1 Six residents and two support workers would like to order curries.

Fill in the table to show 10% of the price of each Balti meal and the discounted cost.

> **Remember**: 10% = 10p in every £ or divide by 10.

		Meat	Prawn	Veg
Balti	10% of price			
	cost			
Dhansak	10% of price			
	cost			
Dopiaza	10% of price			
	cost			
Biryani	10% of price			
	cost			

2 Fill in the table to show 15% of the price of each curry and the discounted cost.

> **Remember**: you can find 10% and then 5%, and then add them together to make 15%.

		Meat	Prawn	Veg
Curry	15% of price			
	cost			
Jalfrezi	15% of price			
	cost			
Tikka	15% of price			
	cost			

3 Explain another way of finding the cost with a 15% reduction.

4 There is a budget of £35. Choose eight curries that will fit in the budget. What do they cost?

L2

1 It's Cara's birthday soon. The support workers decide to use some of the weekly budget to have a takeaway to celebrate. They have three options and plan to collect the meals before 6:30pm:

- Chinese House: the set menu for two, barbecued spare ribs, crispy chicken with beansprouts, sweet and sour pork and two portions of fried rice.
- Balti Corner: meat balti, veg biryani, veg dopiaza, prawn jalfrezi, meat tikka, three nan breads and two portions of steamed rice.
- Bumper Burger House: mega meal deal, a separate medium pizza, 2 wedges and 2 portions of sweetcorn.

What is the cost for the meal from:

a) Chinese House?

b) Balti Corner?

c) Bumper Burger House?

d) Which would you recommend? Explain why.

2 What fraction of the price of the large pizza is:

a) the price of the medium pizza?

b) the price of the small pizza?

3 What fraction of the cost of the mega meal deal is:

a) the family meal deal?

b) the bumper burger deal?

Working with fractions, percentages and decimals

SOURCE Buying toiletries

Sam acts as care worker for an elderly lady who lives alone. She helps to do the shopping. The lady asks her to buy some toiletries. Sam has to find the best value.

B
£2.44
50% extra free

450 ml

Hair conditioner

A
£1.40
12.5% off

250 ml

Hair conditioner

A

Shampoo

300 ml

£3.15
⅓ off

B
£2.10
¼ extra free

Shampoo

250 ml

A
£1.12
Buy one, get one half price

Deodorant
40 ml

B
£1.26
20 ml extra free

Deodorant
60 ml

A
New price
£1.89
Save 66p

TOOTHPASTE
125 ml

B
TOOTHPASTE
100 ml

£1.72
⅓ extra free

B

FOAM BATH

1.2 litre

£0.66
0.2 litres extra free

A

FOAM BATH

500 ml

£0.44
25% off

A
£1.89
12.5% extra free

Talc

225 ml

B

Talc

£2.10
20p off

250 ml

E3

1   Which fractions and percentage offers are the same value as:

a) 0.5?

Percentage [   ]   Fraction [   ]

b) 0.25?

Percentage [   ]   Fraction [   ]

2   What is the cost of the following after the discount?:

a) Shampoo A [   ]

b) Foam bath A [   ]

c) Talcum powder B [   ]

d) Two of deodorant A [   ]

e) From the answer to d) above, find the cost of each bottle of deodorant A if you buy two.

[   ]

3   Which shampoo would you recommend Sam buys as better value? Explain why.

4   Which foam bath would you recommend for better value? Explain why.

5   Which talcum powder would you recommend for better value? Explain why.

6   Sam compares the deodorant by finding the cost of each bottle per 10 ml.

a) How much would 10 ml of bottle A cost? (Use your answer to 2e.)

[   ]

b) How much would 10 ml of bottle B cost? [   ]

c) Which bottle do you think Sam should buy? [   ]

L1

1 a) Place all the fractions mentioned in the information on toiletries in the first column of the table below. Place them in order, starting with the largest fraction.

   b) Fill in the values for the equivalent percentages and decimals.

   c) Fill in the equivalent fraction and decimal values for 12.5%.

Fractions	Percentage equivalents	Decimal equivalents
	12.5%	

2 Sam is asked to buy hair conditioner. Which would you recommend? Explain why.

3 Sam sees another shampoo in a bargain bucket: 150 ml for £1. Which shampoo would you recommend? Explain why.

4 a) Sam thinks toothpaste A holds $1\frac{1}{4}$ times the amount in toothpaste B. Is she right?

   b) How much would $1\frac{1}{4}$ times the toothpaste in B cost?

   c) Which toothpaste is the better value?

L2

1 Toothpaste B holds 100g. Sam thinks if there's $\frac{1}{3}$ extra that means 33.3g is free. The shop assistant says it's 25g that is free. Who is right? Explain.

2 How many ml have been added for free to the following?

   a) Talcum powder A

   b) Shampoo B

   c) Conditioner B

3 A bargain bucket contains these items: conditioner 250ml for £1.25, foam bath 500ml for 30p and talcum powder 200ml for £1.50. Compare the costs of the following items. For each, choose the better value and explain your choice.

   a) Conditioner

   b) Foam bath

   c) Toothpaste

   d) Deodorant

   e) Talcum powder

4 For toothpaste B, write the saving as a fraction, decimal and percentage of the original price.

   a) Fraction

   b) Decimal

   c) Percentage

5 For deodorant B write the amount given for free as a fraction, decimal and percentage of the original amount.

   a) Fraction

   b) Decimal

   c) Percentage

Working with fractions, percentages and decimals

 Saving electricity

At Summerville Care Home, the electricity bill has been increasing. The manager knows there have to be some changes. He finds various pieces of information showing average amounts of electricity used and the cost.

Cooking		Lights		Plugs on standby		Computer	
Oven 5 min	$\frac{1}{6}$ kW	Standard 60-watt bulb for 6 hr	$\frac{1}{3}$ kW	Left on at night, daily	1.5 kW	Desktop 1 hr	$\frac{1}{4}$ kW
Microwave 5 min	$\frac{1}{12}$ kW	Energy-saving 11-watt bulb for 6 hr	$\frac{1}{15}$ kW	Turned off	0 kW	Laptop 1 hr	$\frac{1}{16}$ kW
TV		**Shower**		**Kettle**		**Washing machine**	
42-inch plasma TV on for 1 hr	$\frac{1}{3}$ kW	10 min	1 kW	Full kettle	$\frac{1}{5}$ kW	A+ rated	1 kW
19-inch LCD TV on for 1 hr	$\frac{1}{9}$ kW	4 min	$\frac{1}{2}$ kW	Amount needed	$\frac{1}{40}$ kW	D rated	$\frac{1}{2}$ kW

Average electricity use and cost of appliances		
Appliance	**Watts**	**Pence per hour**
blow-dryer	1400	18.48
CD player	25	0.33
desktop computer	300	3.96
dishwasher	1500	19.8
dryer	3000	39.6
fridge	400	5.28
heater	4000	52.8
iron	660	8.712
kettle	900	11.88
laptop	70	0.924
microwave	1600	21.12
oven	3750	49.5
shower	9000	118.8
storage radiator	1800	23.76
TV	250	3.3
vacuum cleaner	600	7.92
washing machine	550	7.26

Equivalent lightbulbs

Standard bulbs	Energy-saving bulbs
100 watt	23–25 watt
75 watt	20–22 watt
60 watt	11–14 watt
40 watt	8–9 watt
25 watt	5–7 watt

Formula to find the cost of using appliances:

$$\frac{watts \times hours\ used \times unit\ cost}{1000}$$

If your appliance lists amps instead of watts, work the watts out first:

amps × 240 volts = watts

E3

1 How much electricity is saved by having a 4-minute shower rather than a 10-minute shower?

2 Tick the number of minutes in a microwave that use the same kWh electricity as 5 minutes in an oven.

a) $2\frac{1}{2}$ ☐   b) $7\frac{1}{2}$ ☐   c) 10 ☐

3 Tick the options that use the same amount of electricity as one standard 60-watt lightbulb for 6 hours.

a) Five 11-watt energy-saving bulbs for 6 hours   ☐

b) Three 11-watt energy-saving bulbs for 6 hours   ☐

c) One 11-watt energy-saving bulb for 15 hours   ☐

d) One 11-watt energy-saving bulb for 30 hours   ☐

4 How many times would you boil just what you needed in a kettle to use the same amount of electricity as boiling a full kettle?

5 If appliances are turned off rather than being left on standby overnight for a week, how many kWh can be saved?

6 Put the following appliances in order of cost per hour, most expensive first.

a) CD player ☐   b) dishwasher ☐   c) dryer ☐   d) heater ☐   e) oven ☐

7 Round the price per hour to the nearest penny for the following appliances:

a) Dishwasher

b) Dryer

c) Heater

d) Oven

e) TV

f) Shower

8 Choose two ways that staff can save electricity straight away.

L1

How many kW are saved by:

Remember to find the equivalent fractions.

a) boiling just what is needed in a kettle rather than a full kettle? [   ]

b) using a laptop for an hour rather than a desktop computer? [   ]

c) watching a 19-inch LCD TV rather than a 42-inch plasma TV? [   ]

2   Put the following appliances in order of cost per hour, most expensive first.

a) CD player   □   e) heater   □

b) desktop computer   □   f) shower   □

c) dryer   □   g) TV   □

d) fridge   □

3   Round the costs per hour of the following appliances to the nearest whole penny.

a) Desktop computer [   ]

b) Fridge [   ]

4   Round the costs per hour of the following appliances to one decimal place.

a) Microwave [   ]

b) Storage radiator [   ]

c) Kettle [   ]

5   How much is spent per hour on the following?

a) A desktop computer and a TV [   ]

b) A washing machine and a dishwasher [   ]

6   What is the difference in the electricity cost per hour between:

a) an oven and a microwave? [   ]

b) a desktop computer and a laptop? [   ]

7   Choose three appliances that the manager could consider replacing with a different make or model when the time comes. Explain your choices.

L2

1 How much electricity does one energy-saving 11-watt lightbulb use in one hour?

[   ]

2 Put the following appliances in order of cost per hour, most expensive first.

a) CD player ☐   e) heater ☐

b) kettle ☐   f) shower ☐

c) laptop ☐   g) TV ☐

d) fridge ☐   h) desktop computer ☐

3 Round the costs per hour of the following appliances to two decimal places.

a) Iron [   ]   b) Laptop [   ]

4 What is the cost of using the following appliances for 4 hours? Give the answers in £s.

a) Desktop computer [   ]   d) CD player [   ]

b) Laptop [   ]   e) Heater [   ]

c) TV [   ]   f) Storage radiator [   ]

5 What is the difference in cost for 4 hours between using:

a) a laptop and a desktop computer? [   ]

b) a TV and a CD player? [   ]

c) a heater and a storage radiator? [   ]

6 Use the formula for finding the cost of using appliances to complete the table at the rate of 13.2p per kW. Give your answers in pence to 2 decimal places.

Standard lightbulbs	Cost for 1 hour	Energy-saving bulbs	Range of costs for 1 hour
100 watt		23–25 watt	
75 watt		20–22 watt	
60 watt		11–14 watt	
40 watt		8–9 watt	
25 watt		5–7 watt	

7 Each resident in the care home has their own room as well as the use of shared areas. What could the manager ask residents to change in their rooms?

Measures, shape and space

Measuring time

 FOCUS ON Reading the date and time

Read dates in different formats

Watch out for the different ways of writing the date. The order is not always the same.

In Britain the day comes first, then the month:

- 4/1/12 = 4 January 2012
- 1/4/12 = 1 April 2012.

In America the month is written first:

- 1/4/12 = January 4 2012
- 4/1/12 = April 1 2012.

Reading and recording time

am and pm

We use **am** and **pm** to split the day into two halves.

- am starts at midnight, 12am, and goes through the night and morning until midday.
- pm starts at midday, 12pm, and goes though the afternoon and evening until midnight.

24-hour clock

Instead of using am and pm, the hours are numbered up to 24 on the 24-hour clock. For pm times add on 12 hours.

You need to write the time on the 24-hour clock using four digits. Add a leading zero if needed. For example:

- 07:30 = 7:30am

Note that two dots (a colon) separate the hours from the minutes. Some timetables miss the colon out.

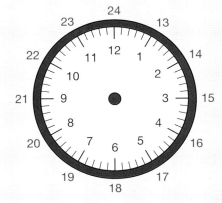

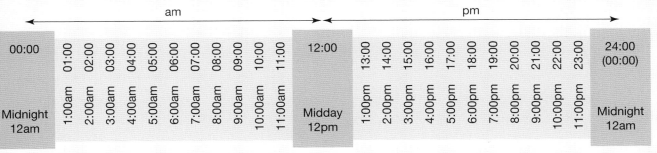

Calculating time

Remember: 60 seconds = 1 minute; 60 minutes = 1 hour

Adding and subtracting hours, minutes and seconds

40 seconds + $1\frac{1}{2}$ minutes	3 hours 24 minutes – 55 minutes
	3 hours 24 minutes = 2 hours + 24 minutes + 60 minutes (1 hour)

minutes	seconds
	40 +
1	30
1	70

70 seconds = 1 minute 10 seconds

1 minute + 1 minute + 10 seconds = 2 minutes 10 seconds

hours	minutes
2	84 –
	55
2	29

2 hours 29 minutes

Calculating lengths of time: adding on
How much time is there between 10:45 and 12:30?

		hours	minutes
Start time	10:45		
Number of minutes to next hour	10:45 → 11:00		15 +
Number of hours	11:00 → 12:00	1	
Number of minutes after the hour	12:00 → 12:30		30
		1	45

Calculating lengths of time: subtracting
How long is it between 8:15am and 2:45pm?

Put into 24-hour clock time, end time first   14:45 –

08:15

6:30 = **6 hours 30 minutes**

Measuring length, weight and capacity

 FOCUS ON Reading scales and measuring

Making quick estimates

You can estimate some measurements quickly by using things you know. You can also use these methods to consider whether an answer you have calculated or measured is reasonable.

- 1 millimetre (mm) is about the width of this full stop (.).
- 1 centimetre (cm) is probably about the width of your little finger.
- 1 metre (m) is probably about the length from your nose to the tips of your fingers.
- 1 gram (g) is about the weight of 10 matchsticks.
- 1 kilogram (kg) is the weight of most bags of sugar.
- 1 litre (l) is the quantity of liquid in most bottles of squash or cartons of juice.

> Check the units on the ruler and always start reading the scale from zero.

Using a ruler

A ruler is a type of scale.
This ruler measures in centimetres (cm) and millimetres (mm).

- 10 mm = 1 cm
- The nail measures 20 mm or 2 cm.

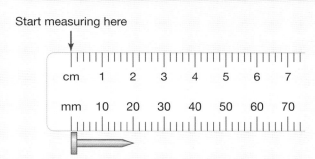

Reading labelled and unlabelled measures

It's easy to read off some measurements on a scale – they are labelled, as on this scale that you might find on a measuring jug.

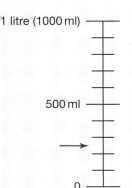

To read off measurements at marked divisions, you need to work out what each division represents. Read two labelled divisions and find the difference. For example:

1000 ml – 500 ml = 500 ml

Divide by the number of divisions from one labelled mark to the next:

500 ml ÷ 5 = 100 ml

On this scale each division is 100 ml.

To read off measurements that are not marked, you need to estimate the distance along the scale between one marked division and the next.

On this scale the arrow points to about halfway between two marked divisions. The unlabelled divisions are 100 ml apart and the arrow points to $2\frac{1}{2}$ divisions up. So the measurement is 250 ml.

Reading measures in decimal format

Look at the position of the numbers and the decimal point (.). Note whether a number is in the tenths column or the hundredths column. For example:

whole
units
decimal point
2.45 cm
tenths
hundredths

Ask yourself: What is the value of one-tenth or one-hundredth of the unit?

> Make sure you always calculate using the same units.

Metres and centimetres

Units	Decimal point	Tenths	Hundredths
1 metre	.	$\frac{1}{10}$ metre	$\frac{1}{100}$ metre
100 cm	.	10 cm	1 cm

Example:

2.45 m
= 200 cm + 40 cm + 5 cm
= 2 m 45 cm or 245 cm

Kilograms and grams

Units	Decimal point	Tenths	Hundredths	Thousandths
1 kg	.	$\frac{1}{10}$ kg	$\frac{1}{100}$ kg	$\frac{1}{1000}$ kg
1000 g	.	100 g	10 g	1 g

You can use the same principle for litres and millilitres.

Calculating measures in decimal format

When adding or subtracting measures with decimals, make sure you line the numbers up with the decimal points under each other.

Example:

1 kilogram + 150 grams + 75 g = 1 . 0 kg
                                0 . 1 5 0 kg
                            +   0 . 0 7 5 kg
                            =   1 . 2 2 5 kg

FOCUS ON Converting measures

Converting between metric measures

To convert between metric measures you need to either multiply or divide by 10, 100 or 1000.

Think about the direction in which you are converting, for example metres to centimetres or centimetres to metres.

Always ask yourself: Would I expect the answer to be larger or smaller than what I started with?

Converting lengths

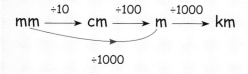

Converting capacity

$$\text{ml} \xrightarrow{\div 1000} \text{litres} \qquad\qquad \text{litres} \xrightarrow{\times 1000} \text{ml}$$

Converting weight

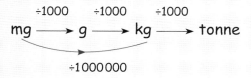

Imperial measures

Remember these quick approximations.

- 1 inch is about the size of a thumb from the knuckle to the end of the nail.

- 1 foot is about the length of many rulers.

- 1 yard is a little shorter than a metre, which is about the length from your nose to the end of your fingers when you stretch your arm out.

- 2 lb (2 pounds) is slightly less than most bags of sugar.

- $1\frac{3}{4}$ pints is the quantity of liquid in most bottles of squash or cartons of juice.

- Most babies are born weighing between $\frac{1}{2}$ and $\frac{3}{4}$ stone.

Converting imperial measures

Converting between imperial measures involves more numbers than with metric measures. You may need to make two calculations.

> **Remember:**
> 12 inches = 1 foot
> 3 feet = 1 yard

Also remember you cannot use a decimal point as you can with metric measures.

Example:

How many yards is 54 inches?

Inches are smaller than yards.
We expect fewer yards, so divide.

54 ÷ 12 (to find feet) = 4 feet, remainder 6 inches

4 ÷ 3 (to find yards) = 1 yard, remainder 1 foot

Answer = 1 yard + 1 foot + 6 inches or 1 yard and 18 inches, but not 1.18 yards.

Converting between metric and imperial measures

Consider whether you need to divide or multiply for the conversion.

Example:

How many inches is 10 cm?

10 ÷ 2.54 = 4 inches

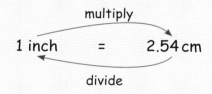

multiply

1 inch   =   2.54 cm

divide

Example:

How many centimetres is 10 inches?

10 × 2.54 = 25.4 cm

> **Remember:**
> - 1 in = 2.54 cm
> - 10 mm = 1 cm
> - 1 oz = 28.35 g
> - 1 lb = 0.454 kg
> - 1000 g = 1 kg

Perimeter, area and scale drawings

 FOCUS ON Measuring areas and volume, and using scale ratios

Area

An area, such as the floor of a room or the expanse of a garden, is measured in square units. The measurement is written with a symbol to show it is squared, for example m² or cm².

The formula is:

Area = length × width or A = l × w

Always use the same units in a calculation, for example:

1.1 m × 0.8 m or 110 cm × 80 cm

When converting from m² to cm², remember to convert the units for both the length and the width measurements.

$$1\,m^2 = \quad 1\,m \begin{array}{c} \xleftrightarrow{\;1\,m\;} \\ \square \end{array} = \quad 100\,cm \begin{array}{c} \xleftrightarrow{\;100\,cm\;} \\ \square \end{array} = 10\,000\,cm^2$$

Perimeter

If you measure the distance all the way round a shape, you have measured the perimeter. For example, the perimeter of a simple shape like a rectangle is two lengths and two widths.

The formula is:

Perimeter = 2 × (length + width) or P = 2 (l + w)

Always use the same units in a calculation, for example:

2 × (110 cm + 60 cm) or 2 × (1.1 m + 0.6 m)

60 cm ↕ ⟷ 1.1 m

Area of a circle

The formula is:

Area = π × radius × radius or A = πr²

> **Remember:** π is often rounded to 3.14, but you can also use the function on your calculator.
>
> Don't confuse πr² and 2πr.
>
> **Remember:** you need *squared* units for area.

Circumference

The perimeter of a circle is called the circumference.

The formula is:

Circumference = π × diameter or **C = πd**

Or **Circumference = 2 × π × radius** or **C = 2πr**

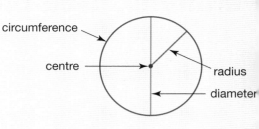

Volume

Volume is measured in cubic units. The measurement is written with a symbol to show it is cubed, for example cm³.

The formula for the volume of a cuboid is:

Volume = length × width × height or **V = l × w × h**

Always use the same units in a calculation.

If converting between units, remember to consider the conversion for all three dimensions.

$1 \, m^3 =$ ⬜ $100 \, cm × 100 \, cm × 100 \, cm = 1\,000\,000 \, cm^3$

> **Remember**: convert volume to litres using 1 litre = 100 cm³.

Using scale ratios

Maps and scale drawings represent real-life measurements at a smaller, specified scale.

For example, a scale of 1:20 on a plan means that the actual measurements are 20 times those on the plan. So if a measurement on a plan is 5 cm, the measurement in reality is 20 times bigger.

5 × 20 = 100 cm (1 m) in reality.

In this case, the ratio of 5 cm to 1 m can be written as 1:20.

To calculate the ratio of 4 cm to 1 km, make the units the same and then simplify the ratio (if possible). For example:

4 cm : 1 km = 4 cm : 1 × 1000 × 100
               = 4 : 100 000
               = 1 : 25 000 (the scale of some Ordnance Survey maps)

> **Remember**: 1 km = 1000 m and 1 m = 100 cm.

Drawing plans and maps

If the scale of a plan or map is 1:*n* (where *n* is a number), divide the actual distances by *n* to give the corresponding distance on the map or plan.

> **Remember**: it is more convenient to work in cm for lengths on a plan or map.

Measuring time

 **MAR sheets for residents at Summerville Care Home**

Tim Lee	May 2012						
	4th	5th	6th	7th	8th	9th	10th
Prednisolone: 5 mg tablets Give 30 mg daily after breakfast for 3 days.							
(0800)	JC	JC	JC				
1200							
1600							
2000							
Date prescribed: 4 May 2012 Signed: Number of tablets in pack: 28							
Amoxicillin: 250 mg tablets Give 2 tablets twice daily. Complete the course of tablets.							
(0800)	/	/	JC	JC			
1200							
1600							
(2000)	/	JC	JC				
Date prescribed: 5 May 2012 Signed: Number of tablets in pack: 32							

E3

Look at the MAR sheet for Tim.

1   On which date does Tim start taking prednisolone?

2   At what time does Tim have the prednisolone?

3   Why should Tim not have the prednisolone in the evening?

4   For how many days does Tim take prednisolone?

5   4 May was a Thursday. On which day of the week did Tim have his last prednisolone?

6   On which date does Tim start taking amoxicillin?

7   Are the statements true or false? Circle the correct response.
   a) Tim will have his next amoxicillin tablet at 8pm.   True/False
   b) There are 8 hours between the times Tim has amoxicillin tablets.   True/False
   c) Altogether Tim has 4 amoxicillin tablets a day.   True/False
   d) Tim has had 16 amoxicillin tablets so far.   True/False
   e) Tim has to take 28 amoxicillin tablets altogether.   True/False

8   On what days and times will Tim take the rest of his amoxicillin tablets?

L1

Look at the MAR sheet for Tim.

1   On which date did Tim start having:
   a) prednisolone?
   b) amoxicillin?

2 Which tablets does Tim take:

a) once a day?

b) twice a day?

3 How many tablets does Tim take each day for:

a) prednisolone?

b) amoxicillin?

4 At what time on which date will Tim have his last amoxicillin tablets?

5 On 11 May the doctor prescribed Ken paracetamol for two days. He is to have eight 500 mg tablets in 24 hours. To be effective, two tablets are taken together at regular intervals. Ken is woken at 6am.

Complete the MAR sheet, showing the dates and times when Ken will be given paracetamol. Circle and sign as if you had given Ken his medication for two days.

Ken	May 2012				
Paracetamol: 500 mg tablets Give 2 tablets 4 times a day at regular intervals for 2 days.					

6 Explain why you chose the above times for Ken to receive his paracetamol.

L2

Mikey comes into Summerville Care Home on 9 May. He has to take amoxicillin. He has to have two 250 mg tablets three times a day at regular intervals. The first tablets are given at 8am and Mikey goes to bed at 10pm.

a) What times will you need to put on Mikey's MAR sheet?

b) There are 30 tablets in the pack. At what time on which date will Mikey have his last tablets?

2 On 13 May the doctor prescribed Jade prednisolone, salbutamol inhaler and erythromycin, as follows:

- 20 mg prednisolone is to be taken once a day (tablets are 5 mg each) for three days.

- One puff of salbutamol inhaler is to be taken four times in every 24 hours for three days, then reduce to one puff three times in every 24 hours for four days.

- Two 250 mg erythromycin tablets are to be taken three times in 24 hours at regular intervals. There are 24 tablets.

Jade has her first tablets at 7:00am and she goes to bed at 11:00pm.

Complete the MAR sheet showing when Jade will be given her prednisolone, salbutamol inhaler and erythromycin. Circle and sign as if you had given Jade her medication for a week.

Jade	May 2012						
Prednisolone: 20 mg (5 mg tablets) Give 4 tablets once a day for 3 days.							
Date prescribed: 13 May Signed: Number of tablets: 12							
Salbutamol inhaler: 1 puff × 4 in 24 h for 3 days 1 puff × 3 in 24 h for 4 days							
Date prescribed: 13 May Signed:							
Erythromycin: 250 mg Give 2 tablets three times in 24 h at regular intervals.							
Date prescribed: 13 May Signed: Number of tablets: 24							

Measuring time

 **SOURCE** Staff rota at Summerville Care Home

Staff at Summerville Care Home work in shifts between 8 o'clock in the morning and 8 o'clock at night.

Week 30 April 2012 to 6 May 2012							
	Mon	**Tues**	**Wed**	**Thurs**	**Fri**	**Sat**	**Sun**
Yasmin	8am–2pm	8am–2pm	7:45am–4:30pm	7:45am–3:15pm			3:45pm–8pm
Ben	2pm–8pm	2pm–8pm	8am–2pm	8am–2pm	2pm–8pm		
Farah	8am–2pm	8am–2pm			8am–2pm	8am–2pm	8am–2pm
Dmitri	2pm–8pm	2pm–8pm			8am–2pm	8am–2pm	8am–2pm
Grace	8am–2pm	8am–2pm	2pm–8pm	2pm–8pm		8am–2pm	2pm–8pm
Nicky			1:30pm–8pm	1:30pm–8pm	2pm–8pm	2pm–8pm	2pm–7pm
Mila			8am–2pm	8am–2pm	8am–2pm	2pm–8pm	8am–2pm

E3

1   Look at the staff rota and answer these questions.

   a) When does a normal early shift start and finish?

   [   ]

   b) When does a normal late shift start and finish?

   [   ]

   c) How many staff are on duty on Sunday for the whole period between 2pm and 4pm?

   [   ]

   d) How many staff are on duty on Wednesday between 1:30pm and 2pm?

   [   ]

2   Which day of the week is 2 May 2012?

   [   ]

3   Dmitri gets into work on Friday at 20 minutes to eight.

   a) Fill in the time on Dmitri's watch.
   b) How long has Dmitri got before his shift starts?

   [   ]

4   Yasmin looks at the clock when she arrives at work on Wednesday.
   Is she early or late?

   [   ]

   07:50

5   The bus outside the home leaves for town at quarter past 2.
   How many minutes does Ben have after work on Wednesday to get to the bus?

   [   ]

6   Complete this pay slip for Ben.

Employee name: Ben Hope			Payroll number: 00632
Week beginning 30/4/12	Day	Hours worked	Number of hours per day
	Mon		
		Total for week	

L1

1 Look at the staff rota for Summerville Care Home.

a) Yasmin phones the home to check when her shift starts on Sunday.

She is told 'quarter to four'. Is that right?

[   ]

b) Yasmin arrives at work on Sunday at five past four.

Write this time in 24-hour clock time.

[   ]

c) How late is she for work?

[   ]

d) She adds the time on at the end of the day. Write the time she finishes in 24-hour clock time.

[   ]

e) After 18:00, the bus to town runs every 20 minutes, at 20 minutes past, 20 minutes to, and on the hour. At what time will Yasmin catch the bus after her late shift on Sunday?

[   ]

2 Nicky writes her work hours on her pay slip in 24-hour clock time.

a) Fill in the hours worked in 24-hour clock time on the pay slip.

b) Fill in the number of hours worked each day.

c) Fill in the total number of hours for the week.

Employee name: Nicky Ward		Payroll number: 00412		
Week beginning 30/4/12	Day	Hours worked	Number of hours per day
	Wed		
	Thurs		
	Fri		
	Sat		
	Sun		
		Total for week	

L1

1 Look at the staff rota for Summerville Care Home. The other staff cover as follows:

- On Wednesday Mila goes off sick for 3 days.
- Dmitri says he will cover on Wednesday for 4 hours until 12pm.
- Grace will start her Wednesday shift 2 hours early.
- On Thursday Farah and Dmitri will do half of the shift each. Farah will start at 8am and Dmitri will do the second half.
- On Friday Nicky will start her shift 4 hours early and Grace will cover 8am–10am.

Rewrite the rota for the week from Wednesday to Sunday in 24-hour clock time.

	Wed	Thurs	Fri	Sat	Sun
Farah					
Dmitri					
Grace					
Nicky					
Mila					

2 Write Yasmin's work hours on her pay slip in 24-hour clock time.

a) Fill in the number of hours worked each day.

b) Fill in the total number of hours for the week.

> **Remember**: you can write the number of hours in decimal format to make calculation easier, e.g. $2\frac{1}{2}$ hours = 2.5 hours.

Employee name: Yasmin Malik		Payroll number: 00326		
Week beginning 30/4/12	**Day**	**Hours worked**		**Number of hours**
	Mon			
	Tues			
	Wed			
	Thurs			
	Sun			
		Total for week		

3 Yasmin is contracted to work 36 hours. How many more/fewer hours should she do?

Measuring time

SOURCE ## Planning meals at Summerville Care Home

Clare helps three residents to cook their lunches. They have two hours to cook, eat and clear up starting from 11:30. The following information has been prepared to help residents calculate how long it will take to prepare and cook different types of food.

Remember: it takes 4 minutes for the oven to heat up.

Preparation

Process	Time
Peeling vegetables or fruit	5 minutes per portion
Making pastry	20 minutes
Making cake mix	12 minutes
Boiling a kettle of water	2 minutes

Cooking using hob, grill and oven

Process	Time
Cooking root vegetables	Boil in a pan of water for 20 minutes.
Cooking frozen mixed vegetables	Boil in a pan of water for 3 minutes.
Stir-frying fresh vegetables	Stir-fry on the hob for 5 minutes.
Cooking chopped meat or mince	12 minutes in a pan, stirring frequently
Cooking homemade beef burger	Fry for 18 minutes in a pan. Turn over halfway through cooking.
Finishing cottage pie	10 minutes to brown under the grill (after all vegetables and mince have been cooked)
Cooking meat casserole (not including preparing the vegetables)	1½ hours in medium preheated oven
Cooking meat chops	5 minutes under high grill, then 10 minutes under low grill. Turn and repeat.
Cooking meat pie with homemade pastry	20 minutes in a high preheated oven
Cooking fish fingers	12 minutes in a high preheated oven
Cooking oven chips	15 minutes in a high preheated oven
Baking small cupcakes	12 minutes in a low preheated oven
Baking a large cake	35 minutes in a low preheated oven
Cooking fruit pie with homemade pastry	20 minutes in a high preheated oven

Cooking vegetables in a microwave

Food (1 portion)	Time
Cauliflower	5½ minutes
Mixed vegetables	4 minutes
Green beans	4½ minutes
Sweetcorn	3 minutes
Casseroled vegetables	9½ minutes
Peas	3¼ minutes

E3

1. One week Clare helps her residents to make a cottage pie using mince, potatoes and carrots.

   Each person prepares their own portion of carrots and potatoes for the pie.

   The mince, all the carrots and all the potatoes are cooked in separate pans at the same time.

   It then takes 5 minutes to put the carrots and mince in the bowl, mash the potato and put it on top of the pie.

   a) Fill in Clare's chart to find the preparation and cooking times.

Process	Minutes
Prepare the vegetables	
Cook the vegetables	
Cook the mince	
Put the ingredients together to make the pie	
Grill	

   b) The vegetables and mince can be cooked at the same time in different pans. Which takes the longest to cook?

   c) What is the total preparation and cooking time?

   d) The session starts at 11:30am. When will they be able to start eating?

   e) How long do they have to eat and clear up?

2. a) How long would it take to make a large cake ready to eat?

   b) They start making a cake at 11:55am. Clare says they'll be able to eat it at quarter to one. Is she right? Explain.

3. How long would it take to cook enough frozen mixed vegetables for Clare and her three residents all together in the microwave?

4 How much time would be saved by cooking the frozen mixed vegetables all together in a pan?

[   ]

L1

1 One day Clare helps her residents make apple and berry pie. They decide on fish fingers, frozen mixed vegetables and chips for the main course.

a) Is the same oven temperature used for each item?

[   ]

b) Why would Clare boil all the mixed vegetables together in a pan rather than putting them all together in a microwave?

[   ]

c) Clare makes a time chart to check all the food is ready at the same time. Complete the chart.

11:30–11:35	11:35							
Prepare apples	Make pastry and put pie together	Preheat oven	Put pie in oven	Put chips in oven	Put fish fingers in oven	Boil kettle	Cook frozen mixed vegetables	Ready to eat

d) How long will they have to eat and clear up? [   ]

2 One day the group have chopped meat with stir-fried vegetables.

Each resident prepares their own portion of carrots, mushrooms, peppers and cabbage.

They stir-fry the vegetables while they are cooking the meat.

a) How long will it take to have the meal ready to eat? [   ]

b) How could they have saved time? [   ]

c) How many minutes could they have saved? [   ]

3 Would Clare be able to help her residents make a meat casserole during a session? Explain your answer.

L2

1 Clare and her three residents cook onions and mince.

Each person prepares their own portion of onion. Then they take 10 minutes to make the mince and onion into burgers. They cook oven chips and frozen peas to have with the burgers.

Complete the time plan for the meal.

11:30									
Prepare onions	Cook mince	Make burgers	Preheat oven	Cook burger on one side	Put chips in oven	Cook burger on other side	Boil kettle	Put peas in pan	Ready to eat

2 Write the preparation and cooking time as a fraction of the total time Clare spends with the residents.

3 A quarter of the remaining time is spent clearing up. Write, as a fraction of the total time Clare spends with the residents:

a) the time spent eating

b) the time spent clearing up.

4 One day the group cooks potatoes, carrots and frozen peas with chops.

Each person prepares their own portion of potatoes and carrots.

a) Draw arrows to show the order of the tasks.

b) Add the missing times to make the shortest possible preparation and cooking time. Some jobs can be done at the same time as preparing the carrots and potatoes.

Process	Order	Start time
Earliest all ready	1	11:30
Cook first side of chops on low	2	
Boil kettle	3	
Cook carrots and potatoes	4	
Prepare carrots and potatoes	5	
Cook second side of chops on high	6	11:45
Cook first side of chops on high	7	11:50
Cook second side of chops on low	8	11:57
Add peas to carrot pan	9	

Measuring length, weight and capacity

SOURCE | Measuring individuals' weight

Three new residents have just arrived at Summerville Care Home. They are weighed on arrival.

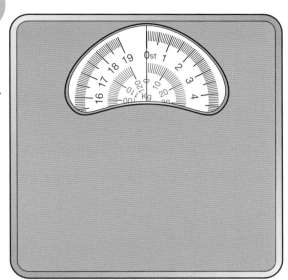

Chris is 6 feet tall.

Tom is 5 feet 10 inches tall.

Marva is 5 feet 2 inches tall.

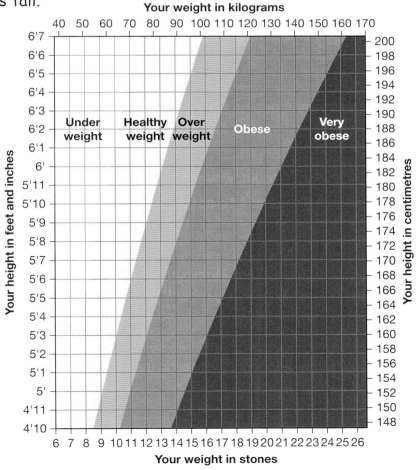

Weight/height chart

E3

1   What does each resident weigh? Fill in the weights in stones and pounds, and also in kilograms.

Resident	Stones and pounds (lb)	Kilograms (kg)
a) Chris		
b) Tom		
c) Marva		

2   Look at the weight/height chart.

a) Which resident is underweight, according to the chart?

b) Which resident is overweight, obese or very obese?

c) Which resident has a healthy weight?

3   The doctor would like Chris to weigh 67 kg.
a) Is that underweight, a healthy weight or overweight?

b) How many more or fewer kilograms is that compared to his present weight?

c) Chris wants to know what his weight would be in stones and pounds.
   Look at the chart to estimate it.

4   The doctor would like Marva to weigh 75 kg.
a) Is that underweight, a healthy weight or overweight?

b) How many kilograms more or less is that compared to her present weight?

c) Marva wants to know what her weight would be in stones and pounds.
   Look at the chart to estimate it.

L1

1 Two weeks later Marva and Chris are weighed again.

What are Chris's and Marva's weights now?

Resident	Stones and pounds (lb)	Kilograms (kg)
a) Chris		
b) Marva		

2 How much has each resident lost or gained? Give the answer in stones and pounds, and kilograms.

a) Chris [ ]

b) Marva [ ]

3 Look at the weight/height chart.

a) Is Chris now underweight, a healthy weight, overweight or obese?

[ ]

b) Is Marva now underweight, a healthy weight, overweight or obese?

Remember: 1 stone = 14 lb.

[ ]

4 The doctor advises that a healthy weight for Chris is 10 stone 7lb or 67 kg.

How much more weight must Chris lose or gain?

Give the answer in stones and pounds, and kilograms.

[ ]

5 The doctor would like Marva to weigh 11 stone 11lb or 75 kg.

How much more weight must Marva lose or gain? Give the answer in stones and pounds, and kilograms.

[ ]

L2

1   Four weeks after arriving at the home Chris and Marva are weighed again.

How much has each resident lost or gained since the first weigh-in? Give the answer in pounds and in kilograms.

a) Chris [   ]

b) Marva [   ]

2   The doctor would like Marva to weigh 11 stone 11 lb or 75 kg.

How many more weeks will this take if Marva loses 3 lb each week?

[   ]

3   Marva wants to lose more to be a healthy weight.

a) Using the weight/height chart, find the highest healthy weight for Marva, in stones and pounds, and in kilograms.

Remember: 1 stone = 14 lb.

[   ]

b) If Marva continues to lose 3 lb a week, how long will it take her to reach a healthy weight?

[   ]

4   The doctor advises that a healthy weight for Chris is 10 stone 7 lb or 67 kg.

a) How many more weeks will this take if Chris gains 2 lb each week?

[   ]

b) Using the weight/height chart, find the lowest healthy weight Chris can be, in stones and pounds, and in kilograms.

[   ]

c) How many more weeks will it take for Chris to reach a healthy weight if he continues to gain 2 lb each week?

[   ]

Measuring length, weight and capacity

SOURCE   Measuring individuals' drinks

Sofia has dementia and is being looked after at home. She finds it difficult to move from her chair.

She has flasks of hot drinks by her chair. One flask contains tea and the other coffee.

She also has a jug of squash by her chair. She pours the squash into a glass whenever she wants.

The tea flask holds 750 ml.

The coffee flask holds 1 litre.

The jug of squash holds 1.5 litres.

The glass holds 250 ml.

Sofia has been advised to drink 2 litres every 24 hours.

Monday:	tea 300 ml; coffee 400 ml;	squash 4 glasses
Tuesday:	tea 250 ml; coffee 750 ml;	squash 3 glasses
Wednesday:	tea 500 ml; coffee 700 ml;	squash 4 glasses
Thursday:	tea 450 ml; coffee 850 ml;	squash 2 glasses
Friday:	tea 300 ml; coffee 450 ml;	squash 5 glasses
Saturday:	tea 650 ml; coffee 650 ml;	squash 3 glasses
Sunday:	tea 525 ml; coffee 375 ml;	squash 4 glasses

E3

1   On which day did Sofia drink most of each drink?

  a) Coffee [   ]

  b) Tea [   ]

  c) Squash [   ]

2   Draw lines to match each day with the total amount that Sofia drank.

Monday	Tuesday	Wednesday	Thursday	Friday	Saturday	Sunday

2200 ml	2000 ml	1700 ml	1900 ml	1800 ml	1750 ml	2050 ml

3   On which day did Sofia drink the recommended amount?

[   ]

4   On which days did Sofia drink more than the recommended amount?

[   ]

5   How much more than the recommended amount did she drink on these days?

[   ]

6   On which day did Sofia did drink the least?

[   ]

7   How much less than the recommended amount was this?

[   ]

L1

1 What is the difference between the most and least that Sofia drank in one day?

┌─────────────────────────────┐
│   │
└─────────────────────────────┘

2 Draw lines to match each day with the total amount that Sofia drank.

Remember: 1 litre = 1000 ml.

| Monday | Tuesday | Wednesday | Thursday | Friday | Saturday | Sunday |

| 1.75 litres | 2 litres | 2.05 litres | 1.8 litres | 1.7 litres | 1.9 litres | 2.2 litres |

3 How many glasses of squash would Sofia have to drink to reach the recommended amount?

┌─────────────────────────────┐
│   │
└─────────────────────────────┘

4 How much more or less than the recommended amount did Sofia drink on:

a) Monday?

b) Tuesday?

c) Wednesday?

d) Thursday?

e) Friday?

f) Saturday?

g) Sunday?

L2

1 How much more or less did Sofia drink:

   a) on Tuesday than on Monday?

   b) on Wednesday than on Tuesday?

   c) on Thursday than on Wednesday?

   d) on Friday than on Thursday?

   e) on Saturday than on Friday?

   f) on Sunday than on Saturday?

2 What fraction of Sofia's drink was tea on:

   a) Friday?

   b) Monday?

3 What fraction of Sofia's drink was coffee on:

   a) Monday?

   b) Wednesday?

4 If Sofia drank four glasses of squash and shared the rest of the recommended amount between tea and coffee, how much tea would she drink?

5 If Sofia drank three glasses of squash and shared the rest of the recommended amount between tea and coffee, how much tea would she drink?

6 One day Sofia drank the recommended amount by drinking all the coffee in the flask and some squash. How many glasses of squash did she drink?

7 On another day Sofia drank the recommended amount by drinking all the tea in the flask and some squash. How many glasses of squash did she drink?

Measuring length, weight and capacity

Abdul recently had an operation. Now he has a catheter, and care staff have to measure fluid output. At first, measurements are taken directly from the catheter bag.

A new bag was fitted at midnight. The bag is emptied every 4 hours.

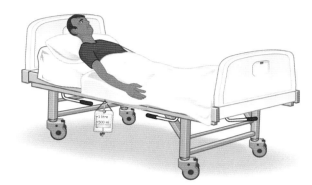

Sunday 5 May

8am
- 2.5 litre
- 2 litre
- 1.5 litre
- 1 litre
- 500 ml

12pm
- 1 litre
- 500 ml

4pm
- 1 litre
- 500 ml

8pm
- 1 litre
- 500 ml

12am
- 1 litre
- 500 ml

The doctor is now very concerned about Abdul.

The catheter is checked and there is nothing wrong with it.

Care staff are asked to take accurate measurements every four hours.

To take accurate measurements, the fluid output has to be weighed.

Monday 6 May

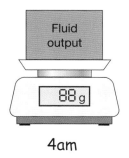
Fluid output
88 g
4am

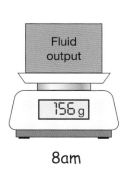
Fluid output
156 g
8am

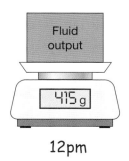
Fluid output
415 g
12pm

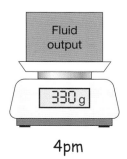
Fluid output
330 g
4pm

E3

1   Complete the chart to show the fluid output from the catheter over the first 24 hours

Daily fluid output balance chart: Abdul		
Sunday, 5 May	Urine volume	Running total
01:00		
02:00		
03:00		
04:00		
05:00		
06:00		
07:00		
08:00		
09:00		
10:00		
11:00		
12:00		
13:00		
14:00		
15:00		
16:00		
17:00		
18:00		
19:00		
20:00		
21:00		
22:00		
23:00		
24:00		

Remember:
100 g of urine
= 100 ml.

2   What measurements are taken on Monday at the following times? Write the answers in millilitres and in grams.

a) 4am

b) 8am

3   a) How often were care staff asked to take accurate measurements of fluid output on Monday?

b) At what time was the fluid output 415 g?

c) At what time was the fluid output 330 g?

d) When will care staff take the next measurement?

L1

1 How much more or less urine output was there on Sunday between:

a) 8am and 12pm?

b) 12pm and 4pm?

c) 4pm and 8pm?

d) 8pm and 12am?

2 How much urine output (in litres) was there altogether on Sunday?

3 How much more or less urine output was there on Monday? Write the answers in millilitres and in grams.

a) Between 4am and 8am

b) Between 8am and 12pm

c) Between 12pm and 4pm

4 How much urine output was there altogether by 4pm on Monday?

5 The next measurement of urine increased by 145g.

a) When was this measurement taken?

b) What was the new measurement (in millilitres and in grams)?

c) What was the new running total for Monday (in litres)?

L2

1. How much urine output was there on Sunday by 8am? [ ]

2. Describe what happened to Abdul's urine output between Sunday 8am and Monday 4pm in terms of increase and decrease. Include the measurements involved.

3. The doctor became concerned at 12pm on Monday. Why do you think this happened?

   [ ]

4. The next urine output (at 8pm) increased by 145g and the final running total for Monday was 1.656 kg.

   Complete Abdul's daily fluid output balance chart for Monday in millilitres or litres.

Daily fluid output balance chart: Abdul		
Monday, 6 May	**Urine volume**	**Running total**
01:00		
02:00		
03:00		
04:00		
05:00		
06:00		
07:00		
08:00		
09:00		
10:00		
11:00		
12:00		
13:00		
14:00		
15:00		
16:00		
17:00		
18:00		
19:00		
20:00		
21:00		
22:00		
23:00		
24:00		

Remember:
100g of urine
= 100 ml.

Measuring length, weight and capacity

SOURCE | Furnishing rooms in care homes

Margo is helping residents to find prices for new curtains and furniture for their rooms. She has listed the measurements and prices of various items.

> The first measurement is the width, the second is the height.

Plain unlined curtains

46 in × 54 in	£12.99
46 in × 72 in	£15.99
54 in × 66 in	£13.99
66 in × 72 in	£21.99
117 cm × 137 cm	£12.99
117 cm × 183 cm	£16.99

Patterned and lined curtains

135 cm × 137 cm	£60
135 cm × 183 cm	£70
135 cm × 229 cm	£80
168 cm × 183 cm	£90
168 cm × 229 cm	£99
228 cm × 229 cm	£130

Chest of drawers

Width:

90 cm	£20
1 m	£25
137 cm	£30

Wardrobes (all 45 cm deep)

Width:

122 cm	£35
137 cm	£42
1.5 m	£50

Bookshelves (all 30 cm deep)

Width:

75 cm	£15
90 cm	£18
137 cm	£22.50

Tables

90 cm × 70 cm	£27.50
90 cm × 90 cm	£32
152 cm × 90 cm	£45

E3

1 Margo draws a plan of how Magdalena would like her room.

 a) Which is the largest-sized wardrobe Magdalena can have?

 b) Which is the largest-sized chest of drawers Magdalena can have?

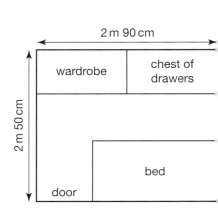

2 Margo measures the window in Magdalena's room. It is 115 cm wide and 125 cm high.

 Magdalena is happy to have unlined curtains.

 Which size of curtains should she buy?

3 How much do Magdalena's wardrobe, chest of drawers and curtains cost, individually and in total?

4 Charlie's window is 110 cm wide and 1.5 m high. He wants patterned lined curtains. Which size of curtains should he buy?

5 Margo draws a plan of how Charlie would like his room.

 Can Charlie have the largest-sized wardrobe and the largest-sized chest of drawers? Explain.

6 How much do Charlie's wardrobe, chest of drawers and curtains cost, individually and in total?

L1

1 The window in Ashok's room is 1.55 m wide and 2 m high. Ashok would like patterned
   lined curtains.

   Which size should he buy?

2 Margo draws a plan of how Ashok would like his room.
   Ashok wants the largest wardrobe and bookshelves
   he can have.

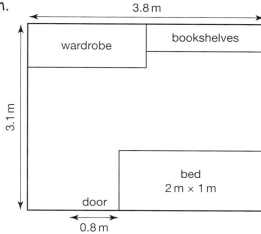

a) What size will they be?

b) What will they cost?

c) How much space will Ashok have left on the wall with the wardrobe and bookshelves?

d) How much space is there between the bed and the bookshelves?

e) Can Ashok fit a table in the space between the bed and the bookshelves? Ashok will
   need 1 m to walk between the bookshelves and the table. Explain your answer.

L2

1   Jason measures his room in feet and inches. Margo
    converts his measurements to inches and then to
    metres and centimetres.

Remember: 1 foot = 12 inches and 1 inch = 2.54 cm.

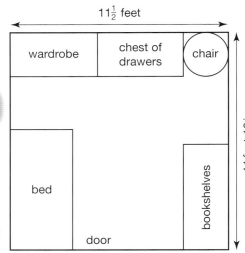

a) Write the dimensions of the room in metres and
   centimetres.

b) Jason already has a circular chair that is 31 inches
   wide. What does it measure in centimetres?

c) Jason wants the largest wardrobe. He then wants the largest chest of drawers he
   can have. Which chest of drawers is this?

d) Jason wants as much space as possible between his chair and the bookshelves.

   Choose bookshelves for Jason. How much free space will there be on this wall?

e) Which tables would be suitable for this space? Explain your answer.

2   Jason wants to know the measurements, but he doesn't understand metres and
    centimetres. Write the width and depth of each of these pieces of furniture in feet
    and inches (to the nearest inch):

a) wardrobe

b) bookshelf

c) possible tables.

3   How much does Jason have to pay for the wardrobe, chest of drawers and bookshelves?

Measuring length, weight and capacity

SOURCE | Posting the mail

At Summerville Care Home, staff help the residents to check the size and weight of their letters and parcels before they go into the mail. They have compiled information about the cost of sending different weights and sizes by first and second class post.

LETTERS		
Weight	**1st class**	**2nd class**
Letter – maximum 240 mm (length) × 165 mm (width) × 5 mm (depth)		
0–100 g	60p	50p
Large letter – maximum 353 mm (length) × 250 mm (width) × 25 mm (depth)		
0–100 g	90p	69p
101–250 g	£1.20	£1.10
251–500 g	£1.60	£1.40
501–750 g	£2.30	£1.90
over 750 g – post as a packet		
Packet – over 353 mm (length) × 250 mm (width) × 25 mm (depth)		
0–750 g	£2.70	£2.20
751–1000 g	£4.30	£3.50
1001–1250 g	£5.60	Items heavier than 1000 g cannot be sent 2nd class.
1251–1500 g	£6.50	
1501–1750 g	£7.40	
1751–2000 g	£8.30	
2001–4000 g	£10.30	
Each additional 2000 g or part of 2000 g	£3.50	

STANDARD PARCELS	
Weight up to:	**Price (all 2nd class)**
2000 g	£5.30
4000 g	£8.80
6000 g	£12.30
8000 g	£15.80
10 000 g	£18.80
20 000 g	£21.90

www.royalmail.com

E3

1 Complete the table below with the weights of each of these items.

A 260 mm × 150 mm × 5 mm

B 200 mm × 150 mm × 3 mm

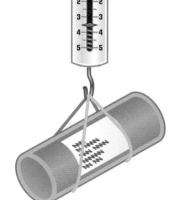

C 400 mm × 150 mm × 150 mm

D 20 cm × 20 cm × 2 cm

E 40 cm × 30 cm × 20 cm

Remember: 10 mm = 1 cm.

Envelope/package	A	B	C	D	E
Weight					

2 Complete this table for each item.

	A	B	C	D	E
Letter, large letter or packet					
Cost for 1st class					
Cost for 2nd class					

3 A postman weighs a parcel on the scales used for C and says it weighs 2 kg 500 g.
   On the scales used for D it says 2 kg 600 g.

   a) Which answer is more accurate?

   b) Explain why.

L1

Complete the table below with the weight and costs of each item.

A 22 cm × 16 cm × 0.5 cm

B 30 cm × 21 cm × 1 cm

C 24 cm × 20 cm × 28 cm

Remember: 10 mm = 1 cm and 1000 g = 1 kg.

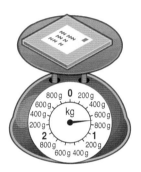

D 20 cm × 20 cm × 2 cm

E 40 cm × 30 cm × 20 cm

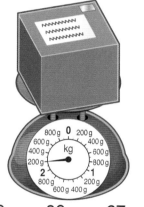

F 30 cm × 30 cm × 27 cm

	A	B	C	D	E	F
Weight (kg)						
Cost 1st class						
Cost 2nd class						

2 Which of the items would be cheaper to send as a parcel rather than as a packet?

3 A resident has three parcels to send to grandchildren at the same address. They weigh 2.75 kg, 2.5 kg and 800 g. Is it cheaper to send them individually or put two or all of them together? Explain your answer.

L2

1 Greta has different-sized envelopes.
Each envelope is $\frac{1}{2}$ in in depth.

Complete the table below for each item.

a) 6 in × 4 in

b) 7 in × 5 in

c) 12 in × 8 in

Remember:
- 1 in = 2.54 cm
- 10 mm = 1 cm
- 1 oz = 28.35 g
- 1 lb = 0.454 kg
- 1000 g = 1 kg

Envelope	Length in mm	Width in mm	1st class	2nd class
a				
b				
c				

2 Greta also has three presents to post. She estimates their weights.

a) 8 oz   b) 2 lb   c) $3\frac{1}{2}$ lb (3 lb 8 oz)

For each present give the weight in grams.

a) [   ]

b) [   ]

c) [   ]

3 For each present, recommend whether Greta should send it as a packet or as a parcel. Explain the possible savings.

a) [   ]

b) [   ]

c) [   ]

Perimeter, area and scale drawings

Designing a new garden for young adults

A supported living organisation buys a new house and carers are told that they can plan a new garden with the young adults.

They plan a lawn area with edging bricks round it and a rectangular flowerbed.

The rest of the garden will be paved with slabs, including 2 m of paving slabs round the outer edges of the lawn and between the lawn and flowerbed.

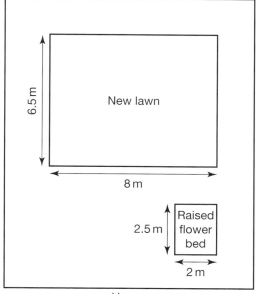

New lawn

6.5 m

8 m

2.5 m | Raised flower bed

2 m

House

E3

1 Edging bricks will go round the lawn.

a) What is the perimeter of the lawn?

b) Each edging brick is 25 cm long.
How many are needed for 1 m?

£1.50 each

25 cm

c) How many edging bricks are needed for the perimeter of the lawn?

d) The bricks cost £1.50 each. What will the total number of bricks cost?

2 A log roll will go round the flowerbed.

a) What is the perimeter of the flowerbed?

b) The rolls are 1.5 m long.
How many are needed to go round the flowerbed?

30 cm x 1.5 m
£10.15

15 cm x 1.5 m
£6.98

c) The flowerbed is to be at least $\frac{1}{4}$ m high.
Which height of log roll should they use?

d) What will all the log roll cost?

3 Turf will be used for the lawn.

1 m x 0.5 m
£1.10 per turf

a) Turf is laid in rows as shown.
How many pieces are used for the length of the lawn?

Turf is laid in rows across the lawn.

b) How many rows are needed?

c) How many pieces of turf are needed for the whole lawn?

d) What is the cost of all the turf?

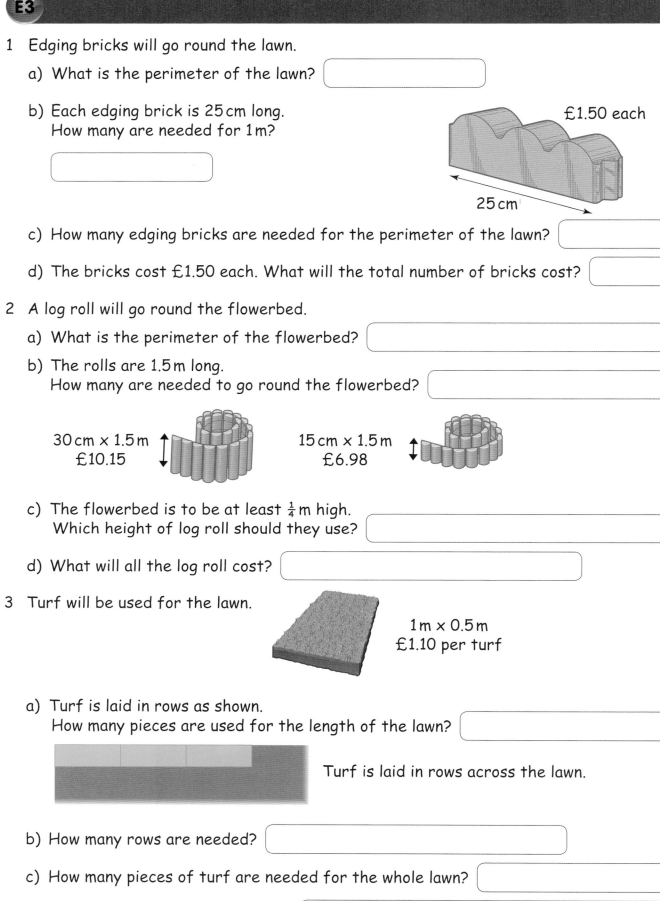

L1

a) What is the area of the lawn?

| |

b) 1kg of grass seed covers approximately 4 square metres.
How much grass seed will be needed?

| |

2 What is the area of the flowerbed?

| |

3 The rest of the garden will have paving slabs.

a) What is the full length and width of the garden?

| |

b) What is the full area of the garden?

| |

c) What is the area of the garden that will have paving slabs?

| |

4 Three sizes of paving slab are available.

50 cm × 50 cm   40 cm × 40 cm   25 cm × 25 cm
£4.49 each   £3.68 each   £2.99 each

Which paving slabs would you recommend? Explain your answer.

L2

1 Log roll is used to edge the flowerbed.

   It is to be at least $\frac{1}{4}$ m high.

   a) Which is the appropriate log roll?

   30 cm x 1.5 m
   £10.15

   b) How much soil will be needed to fill the flowerbed to the top of the logs?

   15 cm x 1.5 m
   £6.98

2 The young adults change part of the plan. They keep the flowerbed but the rest of the garden will be grass apart from a circular area with a diameter of 16 ft for a trampoline. Gravel will be put under the trampoline.

   Safety netting will be needed to go round the circumference of the trampoline.

   a) What is the diameter of the trampoline in metres?

   Remember: 1 ft = 0.3048 m.

   b) What length of netting is required to go round the trampoline?

   c) What is the area of the trampoline?

3 The builder suggests the gravel should be about 2 inches deep.

   Remember:
   1 in = 2.54 cm.

   jumbo bag covers 13 m²
   to a depth of 5 cm

   half a bag covers 6.5 m²
   to a depth of 5 cm

   How many bags of gravel will be needed under the trampoline? Show your working.

4 1 kg of grass seed covers approximately 4 m².
   How many kg of grass seed will be needed to cover the rest of the garden? Show your working.

Perimeter, area and scale drawings

Designing a new dining area

A new care home is opening. The kitchen and dining room need to be planned out.

The staff have measured the dimensions of the dining room as shown below.

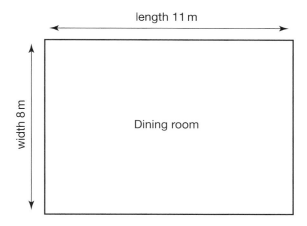

They have also researched the tables available and decided that two options are suitable (shown below).

They have also found out that certain amounts of space are needed for people to be able to sit and walk around comfortably. There needs to be at least 2 m between tables and 1.5 m between a table and a wall.

Square tables
Seat 4 — $1\frac{1}{4}$ m

Rectangular tables
Seat 6 — 2 m × 1 m

E3

1 How many square tables will fit:

   a) along the length of the room? [   ]

   b) across the width of the room? [   ]

   c) One square table seats four people.
      How many people can be seated altogether with square tables?

      [   ]

2 Rectangular tables can be placed widthways or lengthways.

   Widthways   Lengthways   In twos

   If rectangular tables are placed widthways in the room, how many will fit:

   a) along the length of the room? [   ]

   b) across the width of the room? [   ]

   c) One rectangular table seats six people.

      How many people can be seated altogether with rectangular tables placed widthways?

      [   ]

3 If rectangular tables are placed lengthways in the room, how many will fit:

   a) along the length of the room? [   ]

   b) across the width of the room? [   ]

   c) How many people can be seated altogether with rectangular tables placed lengthways?

      [   ]

4 How many rectangular tables will fit if placed in twos:

   a) in the length of the room? [   ]

   b) in the width of the room? [   ]

   c) Two rectangular tables placed together seat eight people.
      How many people can be seated altogether with rectangular tables placed in twos?

      [   ]

5 Which layout would you recommend and why?

L1

1 The dining room is part of a new block. The staff have been given the plan below.

Plan of the new block   Scale: 1 cm to 2 m

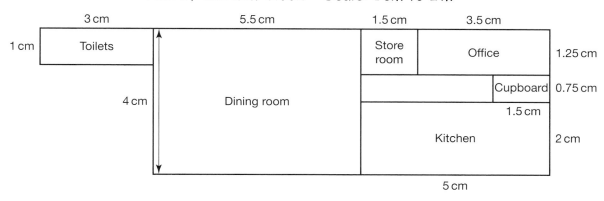

Complete the table with the measurements of the rooms. One of the calculations has been done for you.

Room	Length on plan	Actual length	Width on plan	Actual width
Dining room	5.5 cm	5.5 × 2 = 11 m		
Toilets				
Store room				
Office				
Kitchen				
Cupboard				

2 What size would the different tables be on the plan?

a) square table (to the nearest mm)

b) rectangular table

3 What is the actual length and width of the area that includes the kitchen, store room, office and cupboard?

L2

1   Look at the information on page 97 and the scale drawing of the dining room on page 99.

   If the scale of the drawing was 1:50, what size would the rectangular tables be on the drawing?

   ```
   ┌─────────────────────────────────┐
   │                                 │
   └─────────────────────────────────┘
   ```

2   The residents want to put four rectangular tables (2 m × 1 m) and one cupboard (3 m × 50 cm) into the dining room.

   a) On a separate sheet of graph paper, draw the kitchen using a scale of 1:50. Add rectangles to show where the tables and cupboard could be placed on the drawing. Remember to position them with the required amount of space between the tables, and between the tables and the walls.

   Show your workings below.

   b) Explain your decisions.

3   What is the total area of the dining room?

   ```
   ┌─────────────────────────────────┐
   │                                 │
   └─────────────────────────────────┘
   ```

Handling data

 FOCUS ON Extracting data from tables, charts and graphs

Tables

In order to interpret the data in a table, you need to know what data it includes.

Check the title ──→

Check the row headings. ──→

Check the title and headings for columns. ←──

Number of reported accidents					
	Year				
Site	2008	2009	2010	2011	2012
Brendon	5	4	6	8	4
Darwin	3	0	2	1	1
Fairview	2	4	3	3	2

Charts and graphs

Similarly, you also need to know what data is being presented in a graph or chart. For example, on this bar chart:

Check the title.

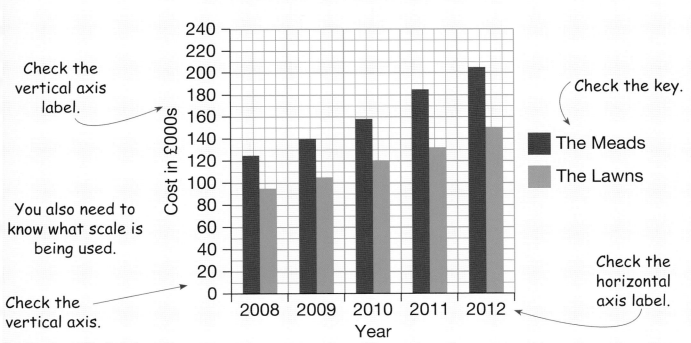

Check the vertical axis label.

Check the key.

You also need to know what scale is being used.

Check the vertical axis.

Check the horizontal axis label.

There are 5 divisions from 0 to 100.

$100 \div 5 = 20$

So each division is worth 20 units, i.e. £20 000.

Note: units are in thousands.

Here's another example of a graph:

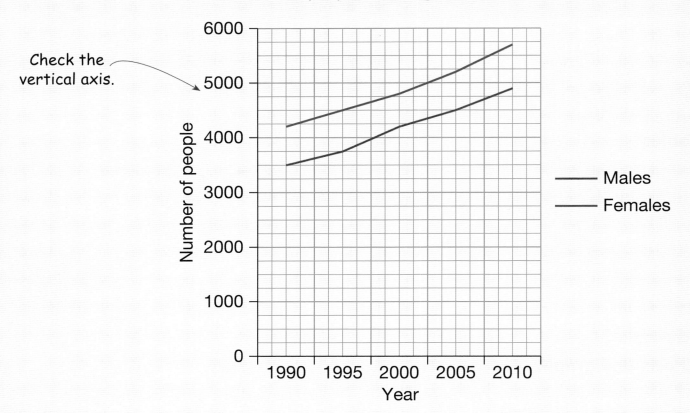

There are 4 divisions from 0 to 1000.

1000 ÷ 4 = 250

So each small division represents 250 units, i.e. 250 people.

FOCUS ON **Presenting data on charts and graphs**

Choosing the right chart or graph

It is important to choose the correct method to display your data. Different charts and graphs are more appropriate for some types of data but not for others.

Bar graph Can be used to display discrete (separate) data or to compare data. For example: • the number of people in different residential homes • the number of males and females in different residential homes.	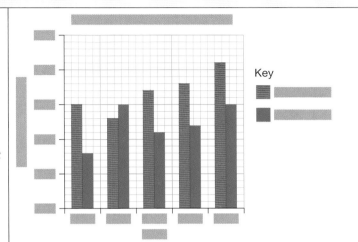
Line graph Can be used to display continuous data or trends. For example: • a patient's temperature • the number of people living to 100 years old.	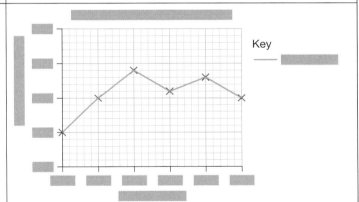
Pie chart Can be used to display proportion. For example: • the make-up of population by ethnicity • a breakdown of local authority expenditure.	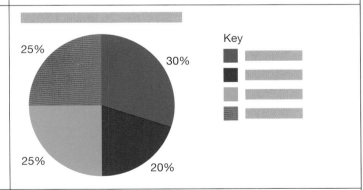
Scattergraph/scattergram Can be used to display correlation (the relationship between two variables). For example: • lifespan vs. alcohol consumption.	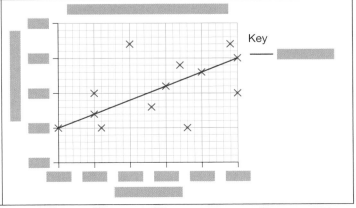

Drawing charts and graphs

Layout

Before you start to draw a graph or bar chart, decide on the largest number you need on the vertical axis. Then decide how many bars or points are needed along the horizontal axis.

Year	1	2	3	4	5	6
Amount (£)	75	118	92	75	60	48

The largest number is 118 so the vertical scale should reach 120.

Count up the large squares and choose a convenient scale, for example 1 large square = £10 or £20. It is usually easiest to count up in multiples of 5 or 10 as each large square has 5 or 10 small divisions on standard graph paper.

Six bars are needed for a bar chart with spaces between the bars. The bars should be the same width as each other. The spaces should be the same width as each other, but can be narrower than the bars.

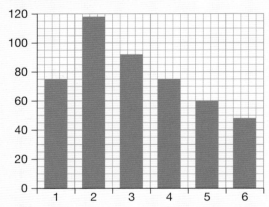

Labelling

All charts and graphs need a title.

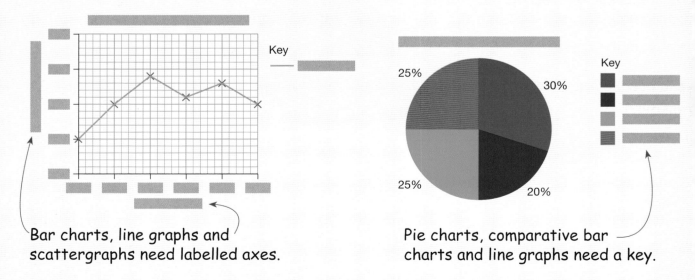

Bar charts, line graphs and scattergraphs need labelled axes.

Pie charts, comparative bar charts and line graphs need a key.

FOCUS ON Averages

An average is a representative value for a set of data. The choice of mean, median or mode depends on the nature of data and what the average is to be used for.

Mean average

The mean is calculated by adding all the values and dividing by the number of values. For example, in a group of people, the ages are:

   17   18   25   17   17   57   65   19   17   18   16

To calculate the mean, add the numbers together = 286

Then divide by the number in the group 286 ÷ 11 = 26

Therefore the mean average age is 26.

Note that the mean may be distorted by extreme values.

Mode

The mode is the most common value.

Using the data above, 17 is the mode as it occurs 4 times (18 occurs twice and the rest only once).

Note that there may be more than one mode or no mode at all.

Median

Put the numbers in order and find the middle number.

Using the data above, put the numbers in order and find the middle number.

   median   16   17   17   17   17   18   18   19   25   57   65

If there is an even number of values, find the midpoint between the two values in the middle. (You can do this by adding them together and dividing by 2.)

Range

The range indicates the spread of the data from the smallest value to the largest value. To calculate the range, subtract the smallest value from the largest.

Using the data above, 65 − 16 = 49.

Note that the smaller the range is, the more consistent the values are.

Extracting data from charts and graphs

SOURCE Investigating obesity

Sylvie is a community health worker who is supporting a group of young mothers. She wants to make them aware of how many children and adults suffer with obesity, and to encourage them to feed their families a healthy diet, so she has collected some data on obesity.

Childhood obesity

The bar chart below shows the number of overweight and obese children in Reception classes (rounded to the nearest 1000) in 2010/11. The data was collected as part of the National Child Measurement Programme.

Adult obesity

This bar chart shows the number of admissions to hospital of obese adults (rounded to the nearest 50).

Overweight and obese Reception children 2010/11

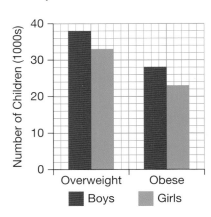

Number of admission episodes with a primary diagnosis of obesity, by Strategic Health Authority, 2010/11

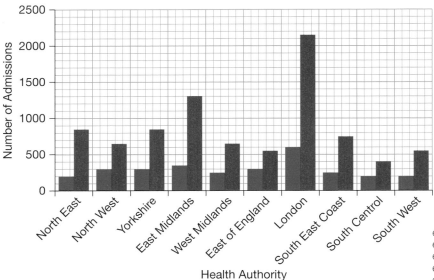

Weight/height chart

> **Remember**: 1 foot = 12 inches and 1 stone = 14 pounds.

The full-size chart is on page 73.

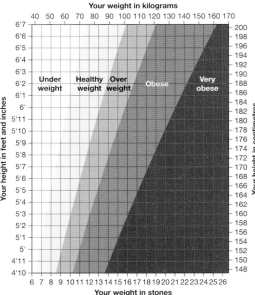

E3

Study the bar chart showing the number of overweight and obese Reception children and answer the following questions.

1   Look at the vertical scale.

a) What does each large division represent?

b) What does each small division represent?

2   How many boys are overweight?

3   How many more boys are overweight than girls?

4   How many girls are obese?

5   What is the difference between the number of obese girls and boys?

6   How many overweight and obese boys are there?

7   How many overweight and obese girls are there?

8   How many overweight and obese children are there?

Now study the bar chart showing hospital admissions of obese adults and answer the following questions.

9   Which health authority has the largest number of female admissions?

10 Which health authority has the largest number of male admissions?

L1

Study the bar chart showing hospital admissions of obese adults and answer the following questions.

1 Which health authorities have 200 male admissions?

2 Which health authorities have 250 male admissions?

3 How many female admissions does the East Midlands have?

4 How many female admissions does London have?

5 What is the difference between the number of male and female admissions for London?

6 How many admissions does Yorkshire have in total?

7 How many more admissions does the East Midlands have in total than the West Midlands?

8 Which health authority has a total of 850 admissions?

L2

Study the weight/height chart on page 73 and answer the following questions.

1 If you are 5' 6" tall and want to be classed as having a healthy weight, what is:

a) the least you can weigh? Give your answer in stones.

b) the most you can weigh? Give your answer in stones.

2 If you are 6' 6" tall and want to be classed as having a healthy weight, what is:

a) the least you can weigh? Give your answer in stones.

b) the most you can weigh? Give your answer in stones.

3 If you are 180 cm tall and weigh 120 kg, what would your weight be classed as?

4 If you are 170 cm tall and weigh 50 kg, what would your weight be classed as?

5 If you are 182 cm tall and weigh 130 kg, approximately how many kilograms do you need to lose to reach a healthy weight?

6 If you are 70" tall and weigh 220 pounds:

a) what would your weight be classed as?

b) approximately how many pounds would you need to lose to reach a healthy weight?

Extracting data from charts and graphs

SOURCE Investigating the ageing population

Philip is a student nurse currently on placement on a ward for older people. Staff who have worked there for some time say that people are living longer. He has researched the following data as he is interested in how this will affect working with elderly people in the future.

Increase in number of centenarians 1981–2001 (rounded to the nearest 100)

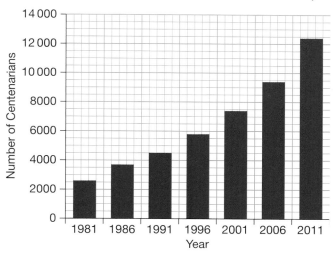

Proportion of men and women over 65 in the UK

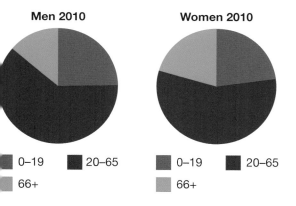

Men 2010   Women 2010

■ 0–19 ■ 20–65 ■ 66+

Projected proportion of men and women over 65 in the UK

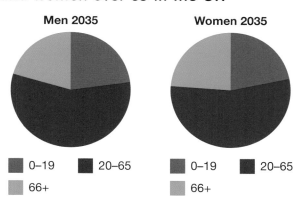

Men 2035   Women 2035

■ 0–19 ■ 20–65 ■ 66+

Year of birth and life expectancy

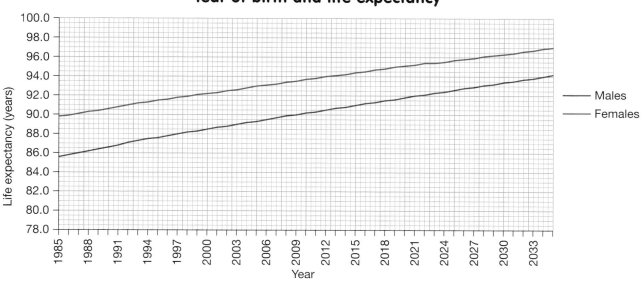

Males
Females

E3

Study the bar chart showing the number of centenarians and answer the following questions.

1   What does each small division on the vertical scale represent?

2   Use the bar chart to estimate how many centenarians there were in:

   a) 1981   e) 2001

   b) 1986   f) 2006

   c) 1991   g) 2011

   d) 1996

3   What is the difference between the number of centenarians in

   a) 1981 and 1986?

   b) 2006 and 2011?

   c) 1981 and 2011?

4   How many centenarians do you think there will be in 2016? Give reasons for your answer.

L1

Study the pie charts showing the proportion of men and women over 65 and answer the following questions.

1   Are these statements true or false?

   a) In 2010 approximately $\frac{1}{4}$ of the male population were aged 0–19.
      True ☐   False ☐

   b) In 2010 the number of men over 65 was approximately twice that of the number aged 0–19.
      True ☐   False ☐

   c) In 2010 over half the female population were aged 20–65.
      True ☐   False ☐

   d) In 2010 the proportion of women over 65 was less than that for men.
      True ☐   False ☐

e) In 2035 the number of men over 65 is predicted to be slightly less than the number aged 0–19.

True ☐   False ☐

f) In 2035 the number of women over 65 is predicted to be slightly more than the number aged 0–19.

True ☐   False ☐

2   Write a sentence summing up the predicted changes to the population according to these pie charts.

L2

Study the line graph for year of birth and life expectancy and answer the following questions.

1   Estimate the life expectancy of:

a) a boy born in 1985

b) a girl born in 1985

c) a boy born in 2012

d) a girl born in 2012

e) a boy born in 2035

f) a girl born in 2035.

2   How much longer can a boy born in 2035 expect to live than one born in 1985?

3   How much longer can a girl born in 2035 expect to live than one born in 1985?

4   Write a sentence summing up life expectancy trends according to this graph.

5   What is your life expectancy according to the graph?

Extracting data from charts and graphs

 Raising awareness on alcohol and health

A health organisation is concerned that people are drinking more than the recommended level of alcohol. They want to use the data below as part of an 'alcohol awareness' campaign.

NHS recommendations on units of alcohol

Men should not regularly* drink more than 3–4 units of alcohol a day

Women should not regularly* drink more than 2–3 units a day

If you've had a heavy drinking session, avoid alcohol for 48 hours

*'Regularly' means drinking this amount every day or most days of the week

Weekly units by age in the UK, 2009

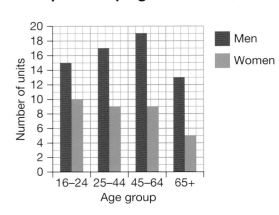

Weekly units by type of drink – proportion of average weekly units accounted for by each type of drink (2009 UK)

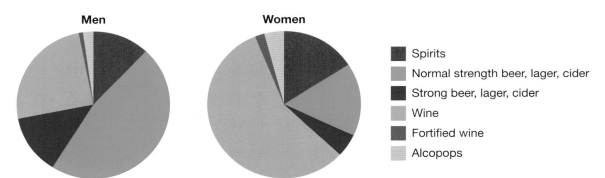

Men   Women

- Spirits
- Normal strength beer, lager, cider
- Strong beer, lager, cider
- Wine
- Fortified wine
- Alcopops

Alcohol-related hospital admissions, England 2009/10

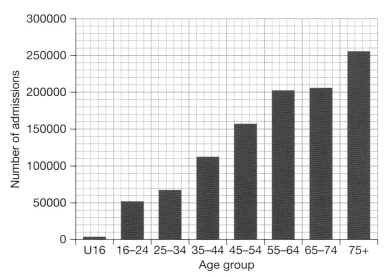

E3

Study the NHS recommendations and the bar chart showing units of alcohol consumed and answer the following questions.

1 Which age group consumes the largest average number of units for:

   a) men?

   b) women?

2 Which age group consumes the smallest average number of units for:

   a) men?

   b) women?

3 What is the difference between the average number of units consumed by men and by women aged:

   a) 45–64?

   b) 16–24?

4 What is the difference between the average number of units consumed by the age groups 16–24 and 65+ for:

   a) men?

   b) women?

5 Use the NHS recommendations to calculate the maximum number of units per week for:

   a) men

   b) women.

6 Do any of the groups of men and women drink more than this amount on average?

L1

Study the NHS recommendations on units of alcohol and answer the following question.

1 If the average number of weekly units for men and women aged 16–24 is below the recommended maximum amount, does this mean that all 16–24-year-olds drink below this amount? Give reasons for your answer.

Now study the two pie charts and answer the following questions.

2   What is the most popular type of drink for:

a) men? [   ]   b) women? [   ]

3   What is the second most popular type of drink for:

a) men? [   ]   b) women? [   ]

4   What is the least popular type of drink for:

a) men? [   ]   b) women? [   ]

5   Decide whether the following statements are true or false.

a) Over half the average weekly units for men are normal strength beer, lager or cider.
   True ☐   False ☐

b) Over half the average weekly units for women are wine.
   True ☐   False ☐

c) Approximately $\frac{1}{8}$ of the average weekly units for men are strong beer, lager or cider.
   True ☐   False ☐

d) Approximately $\frac{1}{4}$ of the average weekly units for women are spirits.
   True ☐   False ☐

L2

Study the bar chart showing alcohol-related hospital admissions and answer the following questions.

1   Use the chart to estimate the number of hospital admissions for each age group.

a) under 16 [   ]   b) 16–24 [   ]

c) 25–34 [   ]   d) 35–44 [   ]

e) 45–54 [   ]   f) 55–64 [   ]

g) 65–74 [   ]   h) 75+ [   ]

2   Use your answer to question 1 to estimate the total number of admissions.

3   Show how you can check your answer to question 2.

4   Write a newspaper headline based on your answer to question 2.

Extracting data from charts and graphs

Investigating adoption rates

Paulina runs a children's home. She is interested in the data below as she is concerned about the likelihood of adoption for the children in her care.

Number of adopted children aged under 1 year in England and Wales, 2000–2011

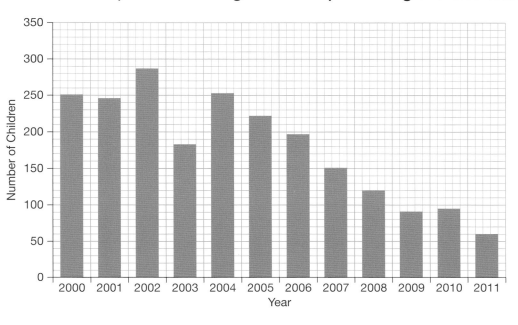

Number of adopted children by age group in England and Wales, 2000–2010

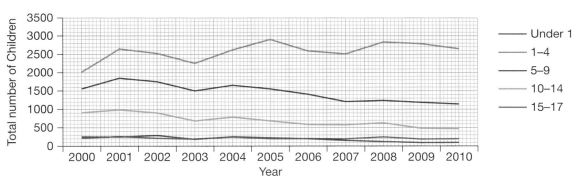

Legend:
- Under 1
- 1–4
- 5–9
- 10–14
- 15–17

Categories of need for adopted children in England and Wales, 2011

- Abuse or neglect
- Child's disability
- Parent's illness or disability
- Family in acute stress
- Family dysfunction
- Socially unacceptable behaviour
- Absent parenting

E3

Study the bar chart showing the number of adopted children aged under 1 year and answer the following questions.

1   In which year were the most children aged under 1 adopted?

[   ]

2   In which year were the fewest children aged under 1 adopted?

[   ]

3   How many children aged under 1 were adopted in:

a) 2000? [   ]

b) 2009? [   ]

4   Estimate the total number of children aged under 1 adopted over the periods:

a) 2000–2002 [   ]

b) 2009–2011. [   ]

5   Comment on what the chart shows about the number of adoptions of children aged under 1.

L1

Study the bar chart showing the number of adopted children by age group and answer the following questions.

1   Which age group had a greater number of adoptions between 2000 and 2010?

[   ]

2   By approximately how many did the number of adoptions for children aged 5–9 decrease between 2000 and 2010?

[   ]

3   By approximately how many did the number of adoptions for children aged 10–14 decrease between 2000 and 2010?

[   ]

4 Estimate the total number of children adopted in

a) 2000

b) 2005

c) 2010.

L2

Study the pie chart showing the categories of need for adopted children and answer the following questions.

1 There were 3050 children adopted in England and Wales in 2011.

Estimate the number of children for the following categories. Justify your estimations.

a) Abuse or neglect

b) Family dysfunction

c) Family in acute stress

d) Absent parenting

e) Parent's illness or disability

f) Child's disability

g) Socially unacceptable behaviour

Presenting data on charts and graphs

SOURCE   Planning a group outing

The staff at a residential home for adults with learning disabilities is planning a trip. They are considering going to a theme park, so they need to check the weather forecast to see which day would be best, and have gathered information about ticket prices, mileage and risk assessment ratings.

Weather forecast

	Morning	Afternoon
Monday	Medium cloud	Medium cloud
Tuesday	Cloudy with sunny intervals	Sunny
Wednesday	Light rain shower	Cloudy with sunny intervals
Thursday	Heavy rain shower	Heavy rain shower
Friday	Medium cloud	Cloudy with sunny intervals

Key　　Medium cloud

Cloudy with sunny intervals

Sunny

Light rain shower

Heavy rain shower

Ticket prices for a theme park

	Minibus group (12–29 people in 1 vehicle)	Coach group (30 or more people in 1 vehicle)
Adult (12+)	£22	£20
Child (4–11)	£22	£20
Older person (65+)	£17	£15
Disabled visitor	£20	£18
Helper for disabled visitor	£20	£18

Free entrance for 1 organiser per coach or minibus!

Mileage chart

Birmingham	Brighton	Bristol	Cambridge	Cardiff	Carlisle	Camarthen	Colchester	Dorchester	Dover	Edinburgh	Exeter	Fort William	Glasgow	Gloucester	Guildford
170															
88	169														
98	120	171													
108	205	48	205												
199	374	282	259	302											
169	266	109	266	68	286										
171	112	195	48	231	310	292									
170	117	62	180	129	364	191	208								
208	81	206	124	241	401	303	116	201							
298	474	381	336	401	99	385	388	463	461						
161	172	83	250	121	355	182	275	55	244	454					
409	585	492	469	512	210	496	520	574	611	133	566				
296	472	379	356	399	97	383	408	461	499	46	454	102			
53	155	36	150	65	247	127	171	118	192	346	110	458	344		
128	44	106	91	142	331	203	103	98	97	431	147	542	428	101	

Risk assessment rating

Severity	Rating	Likelihood	Rating	Risk ranking (likelihood × severity)	
				Value	**Ranking**
Minor injury, e.g. cut or graze	1	Very unlikely	1		
More serious injury, e.g. sprained ankle	2	Unlikely	2	1–5	Low risk
Serious injury, e.g. broken leg	3	Likely	3	6–15	Medium risk
Serious injury affecting more than one person	4	Very likely	4	16–25	High risk
Fatality	5	Certain	5		

E3

Study the weather forecast and answer the following questions.

1 On which day or days might you need sun cream?

2 On which day or days might you need a waterproof coat?

3 Which day do you think would be the best day to visit a theme park?
   Give reasons for your answer.

Now study the ticket prices and answer the following questions.

4 If 15 people travel in a minibus, how much will:

   a) an adult ticket cost?

   b) a ticket for an adult with a disability cost?

   c) a ticket for a helper cost?

   d) How many people can get in free?

5 If 34 people travel in a coach, how much will:

   a) an adult ticket cost?

   b) a ticket for an adult with a disability cost?

   c) a ticket for a helper cost?

   d) How many people can get in free?

6 The residents have been fundraising for the trip. The table shows how much money they have raised from different activities.

Activity	Amount raised (£)
Cake sale	5
Sponsored walk	10
Car wash	12
Plant sale	7
Craft sale	8
Dog walking	9

Create a pictogram that could be displayed for the residents to see how much each activity has raised.

L1

The staff want to know which trip (Theme park, Ice skating, Zoo, Paintballing or Climbing) and which day (Monday–Friday) the residents would prefer.

Design a data collection sheet that the staff could use to record the residents' preferences.

Study the mileage chart and answer the following questions.

The staff are considering lots of options for the trip and need to know how far it is to different locations.

If the home is in Bristol, how far is it to:

a) Gloucester?

b) Exeter?

c) Brighton?

The staff do not want to travel more than 120 miles on the outward trip. What other locations could they choose?

L2

Study the risk assessment ratings and answer the following questions.

1   The staff need to consider the risk ranking for different types of trip.
    Complete the table with the ratings and ranking for each trip.

Trip	Likelihood	Rating	Severity	Rating	Value	Ranking
Theme park	Likely		Minor injury			
Ice skating	Likely		More serious injury			
Paintballing	Very likely		Minor injury			
Zoo	Very unlikely		Minor injury			
Climbing	Unlikely		Serious injury (1 person)			

2   The staff need to calculate and compare the cost of the different trips and present the information to their manager.

    Design a table to record and present this information. The cost of the tickets for residents and staff will need to be calculated, plus the cost of transport and any additional expenses.

Presenting data on charts and graphs

 **SOURCE**   Raising awareness on obesity

Sylvie is a community health worker. She is organising a 'healthy eating' campaign and has collected the information below. Sylvie wants to use this data to create posters to make the public aware of the number of children and adults who are obese.

Table 1 Overweight and obese Year 6 children 2010/11 (rounded to nearest 1000)

	Overweight (1000s)	Obese (1000s)
Boys	36	52
Girls	35	42

Table 2 Admission episodes with a primary diagnosis of obesity by age group, 2010/11 (rounded to nearest 50)

	Under 16	16–24	25–34	35–44	45–54	55–64	65–74	75 and over
2010/11	550	400	1450	3300	3550	1880	450	110

Table 3 Admission episodes with a primary diagnosis of obesity by gender, 2000/01–2010/11

	2000/ 2001	2001/ 2002	2002/ 2003	2003/ 2004	2004/ 2005	2005/ 2006	2006/ 2007	2007/ 2008	2008/ 2009	2009/ 2010	2010/ 2011
Males	309	284	427	498	589	746	1047	1405	2077	2495	2919
Females	741	731	848	1213	1442	1786	2807	3613	5910	8074	8654

E3

1 a) Draw a bar chart using the information in Table 1.

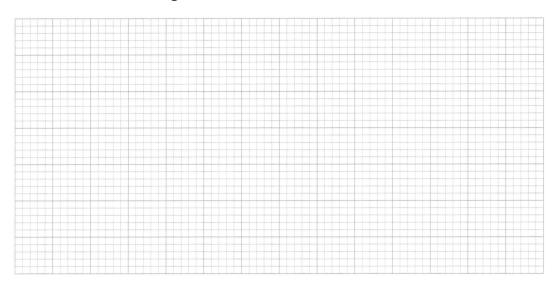

b) Compare your chart with the one for Reception children shown on page 106.

What do you notice about the number of obese and overweight Reception children compared with the number of Year 6 children?

L1

1 a) Draw a bar chart using the information in Table 2.

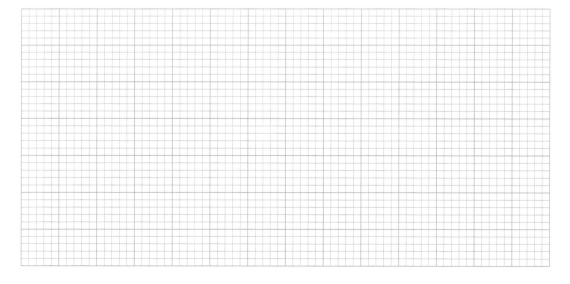

b) Approximately what proportion of the total admissions are 35–54 years old?

L2

a) Draw a line graph comparing the number of male and female admissions from 2000/01 to 2010/11 using the information in Table 3.

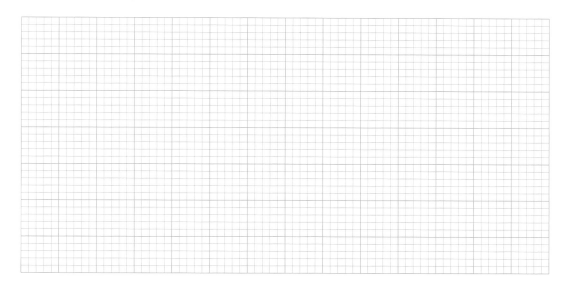

b) Using your graph, comment on the trend for the number of male and female admissions.

Presenting data on charts and graphs

 **Raising awareness on hearing loss**

Following 'Noise Action Week', a local organisation supporting people with hearing loss decides to collect their own data. They want to use this information to produce leaflets to raise awareness of the issues.

Major survey for Noise Action Week (2012) reveals shocking lack of awareness of tinnitus and dangers of loud music

A major survey of the listening habits of the nation reveals most people don't know how much damage loud music can do to their hearing.

Action on Hearing Loss surveyed 1000 people throughout the UK and 83% said they'd suffered from temporary tinnitus and had 'ringing in their ears'

but one in five would only 'be a bit worried' if they got tinnitus permanently.

There is also a general lack of awareness as 80% of people admitted they didn't know loud music can damage their ears or cause tinnitus.

The UK charity is especially concerned about the dangerous

volume levels of people's MP3 players. From next year it will become EU law that all new MP3 players have a maximum default volume setting of 85dB. But the research revealed a staggering 1 in 3 people would override this setting even though this could result in damaging their hearing or developing tinnitus.

Number of people registered as deaf by age, 2010

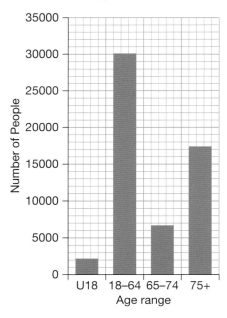

Number of people registered as hard of hearing by age, 2010

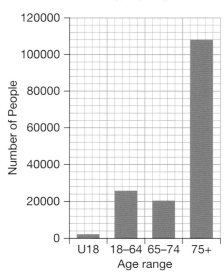

E3

Read the article about Noise Action Week 2012 and answer the following questions.

The survey found that people were not aware of the dangers of listening to loud music.

Create a questionnaire that could be used to find out how aware the people you work or study with are. Use the space below.

L1

Study the bar charts showing the number of people who are deaf and hard of hearing.

1   The charts show the numbers of people by age group.

Design a data collection sheet that could be sent to agencies and organisations to collect data for the current year.

L2

Read the article about Noise Action Week 2012 and answer the following questions.

1   Do the findings fairly represent the views of everyone in the UK? What further information would you need to decide this?

2   What sort of graph or chart would you use to represent the following data?

Draw a rough sketch of what it might look like. Label the axes and/or key.

a) The number of deaf people employed in different occupations.

b) A comparison of the number of registered deaf men and women by age group.

c) The proportion of registered deaf people within each age group.

d) The increase in the number of deaf people over the last 10 years.

e) The connection between the noise level of music listened to on an MP3 player in dB and the level of hearing loss.

Interpreting data from charts and graphs

 **Surveying alcohol consumption**

A survey is carried out with a group of 70 people as part of an 'alcohol awareness' campaign. The number of units of alcohol each person drank on the previous Saturday nigh has been recorded below.

Survey results

0	7	5	2	4	0	3
3	6	9	5	3	3	8
7	3	0	0	1	5	0
4	6	2	4	4	3	5
8	3	2	0	3	4	6
10	7	3	2	3	4	5
0	3	3	2	4	8	4
4	2	6	0	1	3	4
6	5	3	7	2	3	0
3	0	4	2	4	0	5

E3

Record the data in the tally chart below. Count the tallies for each number of units and record the total in the third column.

Number of units of alcohol	Tally	Frequency
0		
1		
2		
3		
4		
5		
6		
7		
8		
9		
10		
	Total	

Remember: check you have not missed any of the data. The total should equal the number of people who took part in the survey.

L1

Use the data in the tally chart to draw a bar chart on the graph paper below. Put the number of units drunk per person along the horizontal axis and the number of people on the vertical axis.

L2

1   Write the frequency for each number of units in the table below and complete the third column.

Number of units of alcohol	Frequency	Number of units × Frequency
0		
1		
2		
3		
4		
5		
6		
7		
8		
9		
10		
	Total	

2   Use this information to calculate the following for the number of units drunk per person

a) the mean

b) the median

c) the mode

d) the range

Interpreting data from charts and graphs

 **SOURCE** Costing residential care

Glen has worked as a care home manager for a number of years. He is part of a committee looking at the current cost of care, the likely cost in years to come and ways that this could be reduced without affecting the quality of care.

Cost of residential care homes		
Average weekly charge by region	2006/07 (£)	2010/11 (£)
Scotland	413	542
Wales	360	468
Northern Home Counties	472	610
Southern Home Counties	454	580
West Midlands	371	470
East Anglia	400	499
North	359	439
Yorkshire	376	447
East Midlands	386	460
South West	420	500
North West	371	435
London	564	597

Source: Laing & Buisson, Age UK

NHS personal social services – gross expenditure (£ million)						
	2004/05	2005/06	2006/07	2007/08	2008/09	2009/10
People aged 65 and over	7970	8390	8660	8770	9080	9390
Adults with physical/learning disabilities and mental health needs (aged 18–64)	5090	5530	5780	6050	6530	6880

E3

Study the table showing the cost of residential care homes and answer the following questions.

1   Fill in the table below, rounding the numbers to the nearest 10.

Residential care homes – average weekly charge by region		
Region	£ 2006/07	£ 2010/11
Scotland		
Wales		
Northern Home Counties		
Southern Home Counties		
West Midlands		
East Anglia		
North		
Yorkshire		
East Midlands		
South West		
North West		
London		

2   Use your rounded figures to complete the comparative bar chart below. Label the axis and give the chart a title.

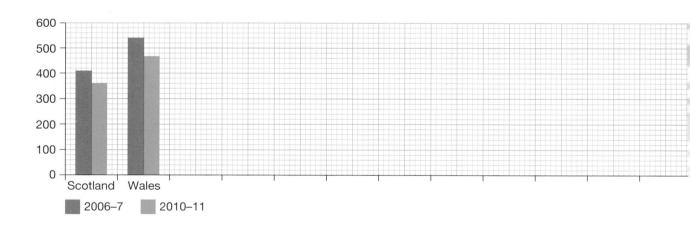

L1

Study the table showing the cost of residential care homes and answer the following questions.

Between 2006/07 and 2010/11, which region had:

a) the biggest increase in the average weekly charge? [   ]

b) the smallest increase in the average weekly charge? [   ]

Using the figures in the table, find the mean charge and range for the regions in:

a) 2006/07

b) 2010/11.

c) What is the difference between the mean averages for 2006/07 and 2010/11?

[   ]

Now study the table showing the cost of NHS personal social services and answer the following questions.

Fill in the table below rounding the numbers to the nearest 100 million.

	2004/05	2005/06	2006/07	2007/08	2008/09	2009/10
People aged 65 and over						
Adults with physical/learning disabilities and mental health needs (aged 18–64)						

Use your table to draw a comparative bar chart on the graph paper.

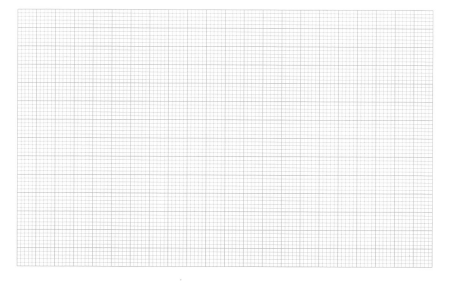

L2

Study the table showing the cost of residential care homes and answer the following questions.

1   Using the figures in the table, find the median charge for the regions in:

a) 2006/07

b) 2010/11.

New College Nottingham
Learning Centres

2   Draw a line graph using the figures for people aged 65 and over.

Make sure you use a suitable scale to enable you to make the predictions in Question 3.

3   Use your line graph to predict what expenditure on over 65s is likely to be in:

a) 2012/13

b) 2014/15.

Give your answers in £ billion.

4   What do you think expenditure for people over 65 will be in 2015? Give reasons for your answer.